# ATI TEAS 6 Subject Test Mathematics

## Student Practice Workbook

## + Two Realistic ATI TEAS Math Tests

Math Notion

www.MathNotion.com

# ATI TEAS 6 Subject Test – Mathematics

# ATI TEAS 6 Subject Test – Mathematics

**ATI TEAS 6 Subject Test Mathematics**

Published in the United State of America By

The Math Notion

Web: WWW.MathNotion.com

Email: info@Mathnotion.com

Copyright © 2021 by the Math Notion. All rights reserved. No part of this publication may be reproduced, stored in a retrieval system, or transmitted in any form or by any means, electronic, mechanical, photocopying, recording, scanning, or otherwise, except as permitted under Section 107 or 108 of the 1976 United States Copyright Ac, without permission of the author.

All inquiries should be addressed to the Math Notion.

ISBN: 978-1-63620-041-5

# ATI TEAS 6 Subject Test – Mathematics

# The Math Notion

**Michael Smith** has been a math instructor for over a decade now. He launched the Math Notion. Since 2006, we have devoted our time to both teaching and developing exceptional math learning materials. As a test prep company, we have worked with thousands of students. We have used the feedback of our students to develop a unique study program that can be used by students to drastically improve their math scores fast and effectively. We have more than a thousand Math learning books including:

- DAT Math Prep
- Accuplacer Math Prep
- Common Core Math Prep
- HiSET Math Prep
- GED Math Prep
- many Math Education Workbooks, Study Guides, Practice and Exercise Books

As an experienced Math test preparation company, we have helped many students raise their standardized test scores—and attend the colleges of their dreams: We tutor online and in person, we teach students in large groups, and we provide training materials and textbooks through our website and through Amazon.

You can contact us via email at:

info@Mathnotion.com

# ATI TEAS 6 Subject Test – Mathematics

## Get the Targeted Practice You Need to Ace the ATI TEAS Math Test!

**ATI TEAS Subject Test - Mathematics** includes easy-to-follow instructions, helpful examples, and plenty of math practice problems to assist students to master each concept, brush up their problem-solving skills, and create confidence.

The ATI TEAS math practice book provides numerous opportunities to evaluate basic skills along with abundant remediation and intervention activities. It is a skill that permits you to quickly master intricate information and produce better leads in less time.

Students can boost their test-taking skills by taking the book's two practice ATI TEAS Math exams. All test questions answered and explained in detail.

**Important Features of the ATI TEAS Math Book:**

- **A complete review** of ATI TEAS math test topics,
- Over 2,500 practice problems covering all topics tested,
- The most important concepts you need to know,
- Clear and concise, easy-to-follow sections,
- Well designed for enhanced learning and interest,
- Hands-on experience with all question types
- **2 full-length practice tests** with detailed answer explanations
- Cost-Effective Pricing

Powerful math exercises to help you avoid traps and pacing yourself to beat the ATI TEAS test. Students will gain valuable experience and raise their confidence by taking math practice tests, learning about test structure, and gaining a deeper understanding of what is tested on the ATI TEAS Math. If ever there was a book to respond to the pressure to increase students' test scores, this is it.

# WWW.MathNotion.COM

… So Much More Online!

- ✓ FREE Math Lessons
- ✓ More Math Learning Books!
- ✓ Mathematics Worksheets
- ✓ Online Math Tutors

For a PDF Version of This Book

Please Visit WWW.MathNotion.com

ATI TEAS 6 Subject Test – Mathematics

# Contents

**Chapter 1 : Integers and Number Theory** ................................................. 11
   Rounding ........................................................................................ 12
   Whole Number Addition and Subtraction ......................................... 13
   Whole Number Multiplication and Division ...................................... 14
   Rounding and Estimates .................................................................. 15
   Adding and Subtracting Integers ...................................................... 16
   Multiplying and Dividing Integers .................................................... 17
   Order of Operations ........................................................................ 18
   Ordering Integers and Numbers ...................................................... 19
   Integers and Absolute Value ............................................................ 20
   Factoring Numbers ......................................................................... 21
   Greatest Common Factor ................................................................ 22
   Least Common Multiple .................................................................. 23
   Answers of Worksheets ................................................................... 24

**Chapter 2 : Fractions and Decimals** .......................................................... 27
   Simplifying Fractions ...................................................................... 28
   Adding and Subtracting Fractions ................................................... 29
   Multiplying and Dividing Fractions .................................................. 30
   Adding and Subtracting Mixed Numbers ......................................... 31
   Multiplying and Dividing Mixed Numbers ........................................ 32
   Adding and Subtracting Decimals ................................................... 33
   Multiplying and Dividing Decimals .................................................. 34
   Comparing Decimals ...................................................................... 35
   Rounding Decimals ........................................................................ 36
   Answers of Worksheets ................................................................... 37

**Chapter 3 : Proportions, Ratios, and Percent** ........................................... 40
   Simplifying Ratios ........................................................................... 41
   Proportional Ratios ......................................................................... 42
   Similarity and Ratios ....................................................................... 43
   Ratio and Rates Word Problems ..................................................... 44

# ATI TEAS 6 Subject Test – Mathematics

Percentage Calculations ........................................................................................... 45
Percent Problems ..................................................................................................... 46
Discount, Tax and Tip .............................................................................................. 47
Percent of Change .................................................................................................... 48
Simple Interest ......................................................................................................... 49
Answers of Worksheets ........................................................................................... 50

## Chapter 4 : Exponents and Radicals Expressions ............................. 53

Multiplication Property of Exponents ..................................................................... 54
Zero and Negative Exponents .................................................................................. 55
Division Property of Exponents ............................................................................... 56
Powers of Products and Quotients ........................................................................... 57
Negative Exponents and Negative Bases ................................................................. 58
Scientific Notation ................................................................................................... 59
Square Roots ............................................................................................................ 60
Simplifying Radical Expressions ............................................................................. 61
Answers of Worksheets ........................................................................................... 62

## Chapter 5 : Algebraic Expressions ........................................................ 65

Simplifying Variable Expressions ............................................................................ 66
Simplifying Polynomial Expressions ....................................................................... 67
Translate Phrases into an Algebraic Statement ........................................................ 68
The Distributive Property ......................................................................................... 69
Evaluating One Variable Expressions ...................................................................... 70
Evaluating Two Variables Expressions .................................................................... 71
Combining like Terms .............................................................................................. 72
Answers of Worksheets ............................................................................................ 73

## Chapter 6 : Equations and Inequalities ................................................ 75

One–Step Equations ................................................................................................. 76
Multi-Step Equations ............................................................................................... 77
Graphing Single–Variable Inequalities .................................................................... 78
One–Step Inequalities .............................................................................................. 79
Multi-Step Inequalities ............................................................................................. 80
Systems of Equations ............................................................................................... 81
Systems of Equations Word Problems ..................................................................... 82

# ATI TEAS 6 Subject Test – Mathematics

Answers of Worksheets ..................................................................................... 83

## Chapter 7 : Linear Functions ............................................................. **87**
Finding Slope .................................................................................................... 88
Graphing Lines Using Line Equation ................................................................. 89
Writing Linear Equations ................................................................................... 90
Graphing Linear Inequalities .............................................................................. 91
Finding Midpoint ............................................................................................... 92
Finding Distance of Two Points ......................................................................... 93
Answers of Worksheets ..................................................................................... 94

## Chapter 8 : Polynomials ................................................................... **97**
Writing Polynomials in Standard Form .............................................................. 98
Simplifying Polynomials .................................................................................... 99
Adding and Subtracting Polynomials ................................................................. 100
Multiplying Monomials ...................................................................................... 101
Multiplying and Dividing Monomials ................................................................ 102
Multiplying a Polynomial and a Monomial ........................................................ 103
Multiplying Binomials ........................................................................................ 104
Factoring Trinomials ........................................................................................... 105
Operations with Polynomials .............................................................................. 106
Answers of Worksheets ....................................................................................... 107

## Chapter 9 : Functions Operations and Quadratic ............................ **111**
Evaluating Function ............................................................................................. 112
Adding and Subtracting Functions ...................................................................... 113
Multiplying and Dividing Functions ................................................................... 114
Composition of Functions ................................................................................... 115
Quadratic Equation .............................................................................................. 116
Solving Quadratic Equations ............................................................................... 117
Quadratic Formula and the Discriminant ............................................................ 118
Graphing Quadratic Functions ............................................................................ 119
Answers of Worksheets ....................................................................................... 120

## Chapter 10 : Geometry and Solid Figures ........................................ **123**
Angles .................................................................................................................. 124
Pythagorean Relationship ................................................................................... 125

# ATI TEAS 6 Subject Test – Mathematics

    Triangles ................................................................................................................. 126

    Polygons ................................................................................................................. 127

    Trapezoids .............................................................................................................. 128

    Circles .................................................................................................................... 129

    Cubes ..................................................................................................................... 130

    Rectangular Prism ................................................................................................. 131

    Cylinder ................................................................................................................. 132

    Pyramids and Cone ............................................................................................... 133

    Answers of Worksheets ........................................................................................ 134

## Chapter 11 : Statistics and Probability ................................................ 137

    Mean and Median ................................................................................................. 138

    Mode and Range ................................................................................................... 139

    Times Series .......................................................................................................... 140

    Stem–and–Leaf Plot .............................................................................................. 141

    Pie Graph .............................................................................................................. 142

    Probability Problems ............................................................................................ 143

    Answers of Worksheets ........................................................................................ 144

## Chapter 12 : ATI TEAS 6 Test Review ................................................. 147

    ATI TEAS 6 Mathematics Practice Tests Answer Sheets ................................... 149

    ATI TEAS 6 Practice Test 1 ................................................................................. 151

    ATI TEAS 6 Practice Test 2 ................................................................................. 165

## Chapter 13 : Answers and Explanations .............................................. 179

    Answer Key ........................................................................................................... 179

    Practice Test 1 ....................................................................................................... 181

    Practice Test 2 ....................................................................................................... 187

ATI TEAS 6 Subject Test – Mathematics

# Chapter 1:
# Integers and Number Theory

**Topics that you'll practice in this chapter:**

- ✓ Rounding
- ✓ Whole Number Addition and Subtraction
- ✓ Whole Number Multiplication and Division
- ✓ Rounding and Estimates
- ✓ Adding and Subtracting Integers
- ✓ Multiplying and Dividing Integers
- ✓ Order of Operations
- ✓ Ordering Integers and Numbers
- ✓ Integers and Absolute Value
- ✓ Factoring Numbers
- ✓ Greatest Common Factor (GCF)
- ✓ Least Common Multiple (LCM)

*"Wherever there is number, there is beauty." –Proclus*

ATI TEAS 6 Subject Test – Mathematics

# Rounding

✍ **Round each number to the nearest ten.**

1) 42 = ____        5) 19 = ____        9) 48 = ____

2) 88 = ____        6) 25 = ____        10) 81 = ____

3) 24 = ____        7) 93 = ____        11) 58 = ____

4) 57 = ____        8) 71 = ____        12) 87 = ____

✍ **Round each number to the nearest hundred.**

13) 198 = ____      17) 321 = ____      21) 580 = ____

14) 387 = ____      18) 433 = ____      22) 868 = ____

15) 816 = ____      19) 579 = ____      23) 480 = ____

16) 101 = ____      20) 825 = ____      24) 287 = ____

✍ **Round each number to the nearest thousand.**

25) 1,382 = ____    29) 9,099 = ____    33) 52,866 = ____

26) 3,420 = ____    30) 22,980 = ____   34) 85,190 = ____

27) 4,254 = ____    31) 45,188 = ____   35) 70,990 = ____

28) 6,861 = ____    32) 16,808 = ____   36) 26,869 = ____

WWW.MathNotion.Com

ATI TEAS 6 Subject Test – Mathematics

# Whole Number Addition and Subtraction

✍ Find the sum or subtract.

1) 1,240 + 658 = _____

2) 3,458 − 544 = _____

3) 2,259 − 752 = _____

4) 1,990 + 324 = _____

5) 3,088 + 229 = _____

6) 2,354 + 1,009 = _____

7) 2,855 + 4,455 = _____

8) 5,112 + 4,004 = _____

9) 4,822 − 2,007 = _____

10) 8,380 − 5,288 = _____

11) 3,227 + 4,150 = _____

12) 7,702 − 4,331 = _____

✍ Find the missing number.

13) 720 + ____ = 1,360

14) 2,115 − ____ = 1,103

15) ____ + 3,105 = 6,200

16) 5,250 − 3,280 = ____

17) 8,020 + ____ = 8,990

18) 7,302 − 4,700 = ____

WWW.MathNotion.Com

ATI TEAS 6 Subject Test – Mathematics

# Whole Number Multiplication and Division

✍ **Calculate each product.**

1) 42 × 13 = _____

2) 70 × 15 = _____

3) 40 × 14 = _____

4) 22 × 20 = _____

5) 110 × 11 = _____

6) 150 × 16 = _____

✍ **Find the missing quotient.**

7) 564 ÷ 6 = _____

8) 270 ÷ 3 = _____

9) 640 ÷ 8 = _____

10) 450 ÷ 9 = _____

11) 112 ÷ 7 = _____

12) 1,260 ÷ 9 = _____

13) 3,000 ÷ 10 = _____

14) 2,400 ÷ 8 = _____

15) 3,200 ÷ 40 = _____

16) 6,300 ÷ 90 = _____

✍ **Calculate each problem.**

17) 400 ÷ 5 = N, N = __

18) 2,500 ÷ 10 = N, N = __

19) N ÷ 3 = 150, N = __

20) 42 × N = 252, N = __

21) 660 ÷ N = 330, N = __

22) N × 8 = 336, N = __

WWW.MathNotion.Com

**ATI TEAS 6 Subject Test – Mathematics**

# Rounding and Estimates

✎ **Estimate the sum by rounding each number to the nearest ten.**

1) $13 + 22 =$ _____

2) $71 + 23 =$ _____

3) $61 + 58 =$ _____

4) $56 + 85 =$ _____

5) $368 + 249 =$ _____

6) $330 + 903 =$ _____

7) $471 + 293 =$ _____

8) $1,950 + 2,655 =$ _____

✎ **Estimate the product by rounding each number to the nearest ten.**

9) $32 \times 71 =$ _____

10) $12 \times 33 =$ _____

11) $31 \times 83 =$ _____

12) $19 \times 11 =$ _____

13) $42 \times 76 =$ _____

14) $63 \times 34 =$ _____

15) $19 \times 31 =$ _____

16) $59 \times 71 =$ _____

✎ **Estimate the sum or product by rounding each number to the nearest ten.**

17) $\begin{array}{r} 29 \\ \times\ 12 \\ \hline \phantom{00} \end{array}$

18) $\begin{array}{r} 37 \\ \times\ 26 \\ \hline \phantom{00} \end{array}$

19) $\begin{array}{r} 48 \\ +\ 82 \\ \hline \phantom{00} \end{array}$

20) $\begin{array}{r} 65 \\ +44 \\ \hline \phantom{00} \end{array}$

21) $\begin{array}{r} 37 \\ \times\ 14 \\ \hline \phantom{00} \end{array}$

22) $\begin{array}{r} 71 \\ +\ 32 \\ \hline \phantom{00} \end{array}$

WWW.MathNotion.Com

# ATI TEAS 6 Subject Test – Mathematics

## Adding and Subtracting Integers

✍ **Find each sum.**

1) $14 + (-6) =$

2) $(-13) + (-20) =$

3) $5 + (-28) =$

4) $50 + (-12) =$

5) $(-7) + (-15) + 3 =$

6) $30 + (-14) + 8 =$

7) $40 + (-10) + (-14) + 17 =$

8) $(-15) + (-20) + 13 + 35 =$

9) $40 + (-20) + (38 - 29) =$

10) $28 + (-12) + (30 - 12) =$

✍ **Find each difference.**

11) $(-18) - (-7) =$

12) $25 - (-14) =$

13) $(-20) - 36 =$

14) $34 - (-19) =$

15) $51 - (30 - 21) =$

16) $17 - (5) - (-24) =$

17) $(35 + 20) - (-46) =$

18) $48 - 16 - (-8) =$

19) $62 - (28 + 17) - (-15) =$

20) $58 - (-23) - (-31) =$

21) $19 - (-8) - (-13) =$

22) $(19 - 24) - (-14) =$

23) $27 - 33 - (-21) =$

24) $58 - (32 + 24) - (-9) =$

25) $36 - (-30) + (-17) =$

26) $27 - (-42) + (-31) =$

ATI TEAS 6 Subject Test – Mathematics

# Multiplying and Dividing Integers

### ✍ Find each product.

1) $(-9) \times (-5) =$

2) $(-3) \times 9 =$

3) $8 \times (-12) =$

4) $(-7) \times (-20) =$

5) $(-3) \times (-5) \times 6 =$

6) $(14 - 3) \times (-8) =$

7) $12 \times (-9) \times (-3) =$

8) $(140 + 10) \times (-2) =$

9) $10 \times (-12 + 8) \times 3 =$

10) $(-8) \times (-5) \times (-10) =$

### ✍ Find each quotient.

11) $42 \div (-7) =$

12) $(-48) \div (-6) =$

13) $(-40) \div (-8) =$

14) $54 \div (-2) =$

15) $152 \div 19 =$

16) $(-144) \div (-12) =$

17) $180 \div (-10) =$

18) $(-312) \div (-12) =$

19) $221 \div (-13) =$

20) $(-126) \div (6) =$

21) $(-161) \div (-7) =$

22) $-266 \div (-14) =$

23) $(-120) \div (-4) =$

24) $270 \div (-18) =$

25) $(-208) \div (-8) =$

26) $(135) \div (-15) =$

WWW.MathNotion.Com

ATI TEAS 6 Subject Test – Mathematics

## Order of Operations

✎ Evaluate each expression.

1) $7 + (5 \times 4) =$

2) $14 - (3 \times 6) =$

3) $(19 \times 4) + 16 =$

4) $(16 - 7) - (8 \times 2) =$

5) $27 + (18 \div 3) =$

6) $(18 \times 8) \div 6 =$

7) $(32 \div 4) \times (-2) =$

8) $(9 \times 4) + (32 - 18) =$

9) $24 + (4 \times 3) + 7 =$

10) $(36 \times 3) \div (2 + 2) =$

11) $(-7) + (12 \times 3) + 11 =$

12) $(8 \times 5) - (24 \div 6) =$

13) $(7 \times 6 \div 3) - (12 + 9) =$

14) $(13 + 5 - 14) \times 3 - 2 =$

15) $(20 - 14 + 30) \times (64 \div 4) =$

16) $32 + (28 - (36 \div 9)) =$

17) $(7 + 6 - 4 - 7) + (15 \div 5) =$

18) $(85 - 20) + (20 - 18 + 7) =$

19) $(20 \times 2) + (14 \times 3) - 22 =$

20) $18 + 5 - (30 \times 3) + 20 =$

WWW.MathNotion.Com

ATI TEAS 6 Subject Test – Mathematics

# Ordering Integers and Numbers

✎ **Order each set of integers from least to greatest.**

1) $8, -10, -5, -3, 4$     ___, ___, ___, ___, ___, ___

2) $-10, -18, 6, 14, 27$     ___, ___, ___, ___, ___, ___

3) $15, -8, -21, 21, -23$     ___, ___, ___, ___, ___, ___

4) $-14, -40, 23, -12, 47$     ___, ___, ___, ___, ___, ___

5) $59, -54, 32, -57, 36$     ___, ___, ___, ___, ___, ___

6) $68, 26, -19, 47, -34$     ___, ___, ___, ___, ___, ___

✎ **Order each set of integers from greatest to least.**

7) $18, 36, -16, -18, -10$     ___, ___, ___, ___, ___, ___

8) $27, 34, -12, -24, 94$     ___, ___, ___, ___, ___, ___

9) $50, -21, -13, 42, -2$     ___, ___, ___, ___, ___, ___

10) $37, 46, -20, -16, 86$     ___, ___, ___, ___, ___, ___

11) $-18, 88, -26, -59, 75$     ___, ___, ___, ___, ___, ___

12) $-65, -30, -25, 3, 14$     ___, ___, ___, ___, ___, ___

WWW.MathNotion.Com

# ATI TEAS 6 Subject Test – Mathematics

## Integers and Absolute Value

✎ **Write absolute value of each number.**

1) $|-2| =$

2) $|-27| =$

3) $|-20| =$

4) $|14| =$

5) $|6| =$

6) $|-55| =$

7) $|16| =$

8) $|2| =$

9) $|54| =$

10) $|-4| =$

11) $|-11|$

12) $|88| =$

13) $|0| =$

14) $|79| =$

15) $|-32| =$

16) $|-17| =$

17) $|42| =$

18) $|-46| =$

19) $|1| =$

20) $|-40| =$

✎ **Evaluate the value.**

21) $|-5| - \frac{|-21|}{7} =$

22) $14 - |3 - 15| - |-4| =$

23) $\frac{|-32|}{4} \times |-4| =$

24) $\frac{|7 \times (-3)|}{7} \times \frac{|-19|}{3} =$

25) $|4 \times (-5)| + \frac{|-40|}{5} =$

26) $\frac{|-45|}{9} \times \frac{|-24|}{12} =$

27) $|-12 + 8| \times \frac{|-7 \times 7|}{7}$

28) $\frac{|-11 \times 2|}{4} \times |-16| =$

WWW.MathNotion.Com

ATI TEAS 6 Subject Test – Mathematics

# Factoring Numbers

✎ **List all positive factors of each number.**

1) 9

2) 16

3) 24

4) 30

5) 26

6) 46

7) 20

8) 68

9) 28

10) 98

11) 14

12) 54

13) 55

14) 18

15) 63

16) 34

17) 50

18) 62

19) 95

20) 64

21) 70

22) 45

23) 22

24) 65

ATI TEAS 6 Subject Test – Mathematics

# Greatest Common Factor

✏ **Find the GCF for each number pair.**

1) 6, 2

2) 4, 5

3) 3, 12

4) 7, 3

5) 5, 10

6) 8, 48

7) 6, 18

8) 9, 15

9) 12, 18

10) 4, 36

11) 6, 10

12) 28, 52

13) 25, 10

14) 22, 24

15) 9, 54

16) 8, 54

17) 42, 14

18) 16, 40

19) 9, 2, 3

20) 5, 15, 10

21) 7, 9, 2

22) 16, 64

23) 30, 48

24) 36, 63

ATI TEAS 6 Subject Test – Mathematics

# Least Common Multiple

✎ **Find the LCM for each number pair.**

1) 6, 9

2) 15, 45

3) 16, 40

4) 12, 36

5) 18, 27

6) 14, 42

7) 6, 30

8) 8, 56

9) 7, 21

10) 8, 20

11) 15, 25

12) 7, 9

13) 4, 11

14) 8, 28

15) 28, 56

16) 40, 50

17) 12, 13

18) 22, 11

19) 36, 20

20) 15, 35

21) 18, 81

22) 30, 54

23) 18, 45

24) 75, 25

WWW.MathNotion.Com

# ATI TEAS 6 Subject Test – Mathematics

# Answers of Worksheets

**Rounding**

| | | | |
|---|---|---|---|
| 1) 40 | 10) 80 | 19) 600 | 28) 7,000 |
| 2) 90 | 11) 60 | 20) 800 | 29) 9,000 |
| 3) 20 | 12) 90 | 21) 600 | 30) 23,000 |
| 4) 60 | 13) 200 | 22) 900 | 31) 45,000 |
| 5) 20 | 14) 400 | 23) 500 | 32) 17,000 |
| 6) 30 | 15) 800 | 24) 300 | 33) 53,000 |
| 7) 90 | 16) 100 | 25) 1,000 | 34) 85,000 |
| 8) 70 | 17) 300 | 26) 3,000 | 35) 71,000 |
| 9) 50 | 18) 400 | 27) 4,000 | 36) 27,000 |

**Whole Number Addition and Subtraction**

| | | |
|---|---|---|
| 1) 1,898 | 7) 7,310 | 13) 640 |
| 2) 2,914 | 8) 9,116 | 14) 1,012 |
| 3) 1,507 | 9) 2,815 | 15) 3,095 |
| 4) 2,314 | 10) 3,092 | 16) 1,970 |
| 5) 3,317 | 11) 7,377 | 17) 970 |
| 6) 3,363 | 12) 3,371 | 18) 2,602 |

**Whole Number Multiplication and Division**

| | | | |
|---|---|---|---|
| 1) 546 | 7) 94 | 13) 300 | 19) 450 |
| 2) 1,050 | 8) 90 | 14) 300 | 20) 6 |
| 3) 560 | 9) 80 | 15) 80 | 21) 2 |
| 4) 440 | 10) 50 | 16) 70 | 22) 42 |
| 5) 1,210 | 11) 16 | 17) 80 | |
| 6) 2,400 | 12) 140 | 18) 250 | |

**Rounding and Estimates**

| | | | |
|---|---|---|---|
| 1) 30 | 7) 760 | 13) 3,200 | 19) 130 |
| 2) 90 | 8) 4,610 | 14) 1,800 | 20) 110 |
| 3) 120 | 9) 2,100 | 15) 600 | 21) 400 |
| 4) 150 | 10) 300 | 16) 4,200 | 22) 100 |
| 5) 620 | 11) 2,400 | 17) 300 | |
| 6) 1,230 | 12) 200 | 18) 1,200 | |

# ATI TEAS 6 Subject Test – Mathematics

**Adding and Subtracting Integers**

1) 8
2) −33
3) −23
4) 38
5) −19
6) 24
7) 33
8) 13
9) 29
10) 34
11) −11
12) 39
13) −56
14) 53
15) 42
16) 36
17) 101
18) 40
19) 32
20) 112
21) 40
22) 9
23) 15
24) 11
25) 49
26) 38

**Multiplying and Dividing Integers**

1) 45
2) −27
3) −96
4) 140
5) 90
6) −88
7) 324
8) −300
9) −120
10) −400
11) −6
12) 8
13) 5
14) −27
15) 8
16) 12
17) −18
18) 26
19) −17
20) −21
21) 23
22) 19
23) 30
24) −15
25) 26
26) −9

**Order of Operations**

1) 27
2) −4
3) 92
4) −7
5) 33
6) 24
7) −16
8) 50
9) 43
10) 27
11) 40
12) 36
13) −7
14) 10
15) 576
16) 56
17) 5
18) 74
19) 60
20) −47

**Ordering Integers and Numbers**

1) −10, −5, −3, 4, 8
2) −18, −10, 6, 14, 27
3) −23, −21, −8, 15, 21
4) −40, −14, −12, 23, 47
5) −57, −54, 32, 36, 59
6) −34, −19, 26, 47, 68
7) 36, 18, −10, −16, −18
8) 94, 34, 27, −12, −24
9) 50, 42, −2, −13, −21
10) 86, 46, 37, −16, −20
11) 88, 75, −18, −26, −59
12) 14, 3, −25, −30, −65

**Integers and Absolute Value**

1) 2
2) 27
3) 20
4) 14

# ATI TEAS 6 Subject Test – Mathematics

5) 6
6) 55
7) 16
8) 2
9) 54
10) 4

11) 11
12) 88
13) 0
14) 79
15) 32
16) 17

17) 42
18) 46
19) 1
20) 40
21) 2
22) −2

23) 32
24) 19
25) 28
26) 10
27) 28
28) 88

## Factoring Numbers

1) 1, 3, 9
2) 1, 2, 4, 8, 16
3) 1, 2, 3, 4, 6, 8, 12, 24
4) 1, 2, 3, 5, 6, 10, 15, 30
5) 1, 2, 13, 26
6) 1, 2, 23, 46
7) 1, 2, 4, 5, 10, 20
8) 1, 2, 4, 17, 34, 68
9) 1, 2, 4, 7, 14, 28

10) 1, 2, 7, 14, 49, 98
11) 1, 2, 7, 14
12) 1, 2, 3, 6, 9, 18, 27, 54
13) 1, 5, 11, 55
14) 1, 2, 3, 6, 9, 18
15) 1, 3, 7, 9, 21, 63
16) 1, 2, 17, 34
17) 1, 2, 5, 10, 25, 50
18) 1, 2, 31, 62

19) 1, 5, 19, 95
20) 1, 2, 4, 8, 16, 32, 64
21) 1, 2, 5, 7, 10, 14, 35, 70
22) 1, 3, 5, 9, 15, 45
23) 1, 2, 11, 22
24) 1, 5, 13, 65

## Greatest Common Factor

1) 2
2) 1
3) 3
4) 1
5) 5
6) 8

7) 6
8) 3
9) 6
10) 4
11) 2
12) 4

13) 5
14) 2
15) 9
16) 2
17) 14
18) 8

19) 1
20) 5
21) 1
22) 16
23) 6
24) 9

## Least Common Multiple

1) 18
2) 45
3) 80
4) 36
5) 54
6) 42

7) 30
8) 56
9) 21
10) 40
11) 75
12) 63

13) 44
14) 56
15) 56
16) 200
17) 156
18) 22

19) 180
20) 105
21) 162
22) 270
23) 90
24) 75

# Chapter 2:
# Fractions and Decimals

**Topics that you'll practice in this chapter:**

- ✓ Simplifying Fractions
- ✓ Adding and Subtracting Fractions
- ✓ Multiplying and Dividing Fractions
- ✓ Adding and Subtract Mixed Numbers
- ✓ Multiplying and Dividing Mixed Numbers
- ✓ Adding and Subtracting Decimals
- ✓ Multiplying and Dividing Decimals
- ✓ Comparing Decimals
- ✓ Rounding Decimals

*"A Man is like a fraction whose numerator is what he is and whose denominator is what he thinks of himself. The larger the denominator, the smaller the fraction." –Tolstoy*

ATI TEAS 6 Subject Test – Mathematics

# Simplifying Fractions

✎ Simplify each fraction to its lowest terms.

1) $\frac{5}{10} =$

2) $\frac{28}{35} =$

3) $\frac{27}{36} =$

4) $\frac{40}{80} =$

5) $\frac{14}{56} =$

6) $\frac{32}{48} =$

7) $\frac{52}{65} =$

8) $\frac{15}{60} =$

9) $\frac{80}{160} =$

10) $\frac{55}{77} =$

11) $\frac{28}{112} =$

12) $\frac{32}{64} =$

13) $\frac{63}{72} =$

14) $\frac{81}{90} =$

15) $\frac{35}{105} =$

16) $\frac{25}{70} =$

17) $\frac{80}{280} =$

18) $\frac{12}{81} =$

19) $\frac{36}{186} =$

20) $\frac{240}{540} =$

21) $\frac{70}{560} =$

✎ Find the answer for each problem.

22) Which of the following fractions equal to $\frac{3}{4}$? ____

A. $\frac{60}{90}$   B. $\frac{43}{104}$   C. $\frac{48}{64}$   D. $\frac{150}{300}$

23) Which of the following fractions equal to $\frac{5}{8}$? ____

A. $\frac{125}{200}$   B. $\frac{115}{200}$   C. $\frac{50}{100}$   D. $\frac{30}{90}$

24) Which of the following fractions equal to $\frac{3}{7}$? ____

A. $\frac{58}{116}$   B. $\frac{54}{126}$   C. $\frac{270}{167}$   D. $\frac{42}{63}$

WWW.MathNotion.Com

ATI TEAS 6 Subject Test – Mathematics

## Adding and Subtracting Fractions

🍃 **Find the sum.**

1) $\frac{5}{9} + \frac{4}{9} =$

2) $\frac{1}{2} + \frac{1}{7} =$

3) $\frac{3}{8} + \frac{1}{4} =$

4) $\frac{3}{5} + \frac{1}{2} =$

5) $\frac{1}{4} + \frac{3}{5} =$

6) $\frac{7}{8} + \frac{3}{8} =$

7) $\frac{1}{2} + \frac{7}{10} =$

8) $\frac{2}{5} + \frac{2}{3} =$

9) $\frac{5}{7} + \frac{2}{3} =$

10) $\frac{7}{12} + \frac{3}{4} =$

11) $\frac{5}{6} + \frac{2}{5} =$

12) $\frac{1}{12} + \frac{2}{3} =$

🍃 **Find the difference.**

13) $\frac{1}{3} - \frac{1}{6} =$

14) $\frac{3}{4} - \frac{1}{8} =$

15) $\frac{1}{2} - \frac{1}{3} =$

16) $\frac{1}{4} - \frac{1}{5} =$

17) $\frac{5}{8} - \frac{2}{3} =$

18) $\frac{1}{4} - \frac{1}{7} =$

19) $\frac{5}{6} - \frac{1}{9} =$

20) $\frac{3}{4} - \frac{1}{6} =$

21) $\frac{7}{8} - \frac{1}{12} =$

22) $\frac{8}{15} - \frac{3}{5} =$

23) $\frac{3}{12} - \frac{1}{14} =$

24) $\frac{10}{13} - \frac{7}{26} =$

25) $\frac{6}{7} - \frac{3}{4} =$

26) $\frac{4}{5} - \frac{1}{8} =$

27) $\frac{4}{7} - \frac{2}{35} =$

28) $\frac{9}{16} - \frac{2}{8} =$

29) $\frac{8}{9} - \frac{7}{18} =$

30) $\frac{1}{2} - \frac{4}{9} =$

ATI TEAS 6 Subject Test – Mathematics

# Multiplying and Dividing Fractions

✏️ Find the value of each expression in lowest terms.

1) $\dfrac{1}{5} \times \dfrac{15}{5} =$

2) $\dfrac{9}{12} \times \dfrac{4}{9} =$

3) $\dfrac{1}{16} \times \dfrac{8}{10} =$

4) $\dfrac{1}{24} \times \dfrac{8}{10} =$

5) $\dfrac{1}{5} \times \dfrac{1}{4} =$

6) $\dfrac{7}{9} \times \dfrac{1}{7} =$

7) $\dfrac{6}{7} \times \dfrac{1}{3} =$

8) $\dfrac{2}{8} \times \dfrac{2}{8} =$

9) $\dfrac{5}{8} \times \dfrac{3}{5} =$

10) $\dfrac{4}{7} \times \dfrac{1}{8} =$

11) $\dfrac{7}{15} \times \dfrac{5}{7} =$

12) $\dfrac{3}{10} \times \dfrac{5}{9} =$

✏️ Find the value of each expression in lowest terms.

13) $\dfrac{1}{4} \div \dfrac{1}{8} =$

14) $\dfrac{1}{10} \div \dfrac{1}{5} =$

15) $\dfrac{3}{4} \div \dfrac{1}{5} =$

16) $\dfrac{1}{3} \div \dfrac{5}{6} =$

17) $\dfrac{1}{7} \div \dfrac{8}{42} =$

18) $\dfrac{3}{4} \div \dfrac{1}{6} =$

19) $\dfrac{2}{7} \div \dfrac{7}{13} =$

20) $\dfrac{1}{24} \div \dfrac{3}{16} =$

21) $\dfrac{7}{12} \div \dfrac{5}{6} =$

22) $\dfrac{22}{18} \div \dfrac{11}{9} =$

23) $\dfrac{9}{35} \div \dfrac{3}{7} =$

24) $\dfrac{2}{7} \div \dfrac{8}{21} =$

25) $\dfrac{1}{9} \div \dfrac{2}{5} =$

26) $\dfrac{5}{12} \div \dfrac{3}{5} =$

27) $\dfrac{3}{20} \div \dfrac{1}{6} =$

28) $\dfrac{8}{20} \div \dfrac{3}{4} =$

29) $\dfrac{5}{6} \div \dfrac{2}{9} =$

30) $\dfrac{5}{11} \div \dfrac{3}{4} =$

ATI TEAS 6 Subject Test – Mathematics

# Adding and Subtracting Mixed Numbers

✎ Find the sum.

1) $3\frac{1}{3} + 2\frac{1}{6} =$

2) $4\frac{1}{2} + 3\frac{1}{2} =$

3) $3\frac{3}{8} + 1\frac{1}{8} =$

4) $2\frac{1}{4} + 2\frac{1}{3} =$

5) $3\frac{5}{6} + 2\frac{7}{12} =$

6) $5\frac{4}{15} + 3\frac{3}{5} =$

7) $2\frac{1}{3} + 4\frac{3}{7} =$

8) $3\frac{1}{2} + 4\frac{2}{5} =$

9) $5\frac{2}{5} + 6\frac{3}{7} =$

10) $8\frac{5}{16} + 6\frac{1}{12} =$

✎ Find the difference.

11) $3\frac{1}{4} - 1\frac{3}{4} =$

12) $6\frac{3}{5} - 4\frac{2}{5} =$

13) $4\frac{1}{3} - 3\frac{1}{9} =$

14) $7\frac{1}{7} - 5\frac{1}{2} =$

15) $5\frac{1}{3} - 2\frac{1}{12} =$

16) $8\frac{1}{5} - 4\frac{1}{3} =$

17) $9\frac{1}{4} - 6\frac{1}{8} =$

18) $11\frac{7}{15} - 8\frac{3}{5} =$

19) $14\frac{5}{6} - 11\frac{3}{5} =$

20) $18\frac{2}{7} - 14\frac{1}{5} =$

21) $9\frac{1}{3} - 4\frac{1}{4} =$

22) $6\frac{1}{8} - 4\frac{1}{16} =$

23) $19\frac{3}{8} - 15\frac{1}{3} =$

24) $11\frac{1}{9} - 8\frac{1}{8} =$

25) $17\frac{1}{7} - 11\frac{1}{5} =$

26) $16\frac{2}{9} - 9\frac{5}{7} =$

# ATI TEAS 6 Subject Test – Mathematics

## Multiplying and Dividing Mixed Numbers

✎ **Find the product.**

1) $5\frac{1}{2} \times 2\frac{1}{4} =$

2) $5\frac{1}{3} \times 4\frac{1}{3} =$

3) $5\frac{3}{4} \times 6\frac{1}{4} =$

4) $3\frac{1}{3} \times 2\frac{3}{5} =$

5) $4\frac{8}{10} \times 1\frac{1}{24} =$

6) $6\frac{2}{7} \times 1\frac{1}{11} =$

7) $8\frac{2}{3} \times 3\frac{1}{2} =$

8) $3\frac{4}{7} \times 2\frac{1}{5} =$

9) $5\frac{2}{8} \times 4\frac{1}{6} =$

10) $7\frac{3}{3} \times 1\frac{3}{8} =$

✎ **Find the quotient.**

11) $2\frac{2}{5} \div 4\frac{1}{5} =$

12) $4\frac{1}{6} \div 3\frac{1}{3} =$

13) $6\frac{1}{3} \div 1\frac{1}{2} =$

14) $7\frac{1}{10} \div 2\frac{2}{5} =$

15) $3\frac{1}{3} \div 1\frac{1}{9} =$

16) $1\frac{1}{10} \div 4\frac{1}{2} =$

17) $1\frac{3}{16} \div 5\frac{1}{4} =$

18) $4\frac{1}{3} \div 4\frac{3}{4} =$

19) $9\frac{1}{3} \div 2\frac{1}{4} =$

20) $15\frac{1}{3} \div 5\frac{1}{2} =$

21) $4\frac{1}{6} \div 1\frac{1}{5} =$

22) $1\frac{1}{18} \div 1\frac{2}{9} =$

23) $4\frac{2}{7} \div 1\frac{3}{10} =$

24) $7\frac{1}{3} \div 2\frac{2}{11} =$

25) $8\frac{2}{5} \div 1\frac{1}{6} =$

26) $9\frac{1}{3} \div 2\frac{1}{7} =$

WWW.MathNotion.Com

ATI TEAS 6 Subject Test – Mathematics

# Adding and Subtracting Decimals

✎ Add and subtract decimals.

1) 35.19 − 24.28

2) 34.29 + 42.58

3) 61.20 + 33.75

4) 38.72 − 21.68

5) 57.39 + 26.54

6) 70.24 − 42.35

7) 86.09 − 35.14

8) 54.51 + 32.66

9) 114.21 − 88.69

✎ Find the missing number.

10) ___ + 2.8 = 5.4

11) 4.1 + ___ = 5.88

12) 6.45 + ___ = 8

13) 7.25 − ___ = 3.40

14) ___ − 2.35 = 4.25

15) ___ − 19.85 = 6.54

16) 22.15 + ___ = 28.95

17) ___ − 37.16 = 9.42

18) ___ + 24.50 = 34.19

19) 72.40 + ___ = 125.20

WWW.MathNotion.Com

# ATI TEAS 6 Subject Test – Mathematics

## Multiplying and Dividing Decimals

✎ **Find the product.**

1) $0.5 \times 0.6 =$

2) $3.3 \times 0.4 =$

3) $1.28 \times 0.5 =$

4) $0.35 \times 0.6 =$

5) $1.85 \times 0.6 =$

6) $0.24 \times 0.5 =$

7) $5.25 \times 1.4 =$

8) $18.5 \times 4.6 =$

9) $15.4 \times 6.8 =$

10) $19.5 \times 2.6 =$

11) $32.2 \times 1.5 =$

12) $78.4 \times 4.5 =$

✎ **Find the quotient.**

13) $1.85 \div 10 =$

14) $74.6 \div 100 =$

15) $3.6 \div 3 =$

16) $9.6 \div 0.4 =$

17) $15.5 \div 0.5 =$

18) $32.8 \div 0.2 =$

19) $22.15 \div 1{,}000 =$

20) $53.55 \div 0.7 =$

21) $322.2 \div 0.2 =$

22) $50.67 \div 0.18 =$

23) $77.4 \div 0.8 =$

24) $27.93 \div 0.03 =$

# Comparing Decimals

✎ Write the correct comparison symbol (>, < or =).

1) 0.70 ☐ 0.070

2) 0.049 ☐ 0.49

3) 5.090 ☐ 5.09

4) 2.57 ☐ 2.05

5) 9.03 ☐ 0.930

6) 6.06 ☐ 6.6

7) 7.02 ☐ 7.020

8) 3.04 ☐ 3.2

9) 3.61 ☐ 3.245

10) 0.986 ☐ 0.0986

11) 17.24 ☐ 17.240

12) 0.759 ☐ 0.81

13) 9.040 ☐ 9.40

14) 5.73 ☐ 5.213

15) 9.44 ☐ 9.404

16) 7.17 ☐ 7.170

17) 4.85 ☐ 4.085

18) 9.041 ☐ 9.40

19) 3.033 ☐ 3.030

20) 4.97 ☐ 4.970

ATI TEAS 6 Subject Test – Mathematics

# Rounding Decimals

✎ **Round each decimal to the nearest whole number.**

1) 28.12  3) 16.22  5) 7.95

2) 6.9   4) 8.5   6) 52.7

✎ **Round each decimal to the nearest tenth.**

7) 31.761  9) 94.729  11) 13.219

8) 14.421  10) 77.89  12) 59.89

✎ **Round each decimal to the nearest hundredth.**

13) 8.428  15) 55.3786  17) 62.241

14) 23.812  16) 231.912  18) 19.447

✎ **Round each decimal to the nearest thousandth.**

19) 15.54324  21) 243.8652  23) 67.1983

20) 34.62586  22) 80.4529  24) 72.36788

# ATI TEAS 6 Subject Test – Mathematics

## Answers of Worksheets

**Simplifying Fractions**

1) $\frac{1}{2}$
2) $\frac{4}{5}$
3) $\frac{3}{4}$
4) $\frac{1}{2}$
5) $\frac{1}{4}$
6) $\frac{2}{3}$
7) $\frac{4}{5}$
8) $\frac{1}{4}$
9) $\frac{1}{2}$
10) $\frac{5}{7}$
11) $\frac{1}{4}$
12) $\frac{1}{2}$
13) $\frac{7}{8}$
14) $\frac{9}{10}$
15) $\frac{1}{3}$
16) $\frac{5}{14}$
17) $\frac{2}{7}$
18) $\frac{4}{27}$
19) $\frac{6}{31}$
20) $\frac{4}{9}$
21) $\frac{1}{8}$
22) C
23) A
24) B

**Adding and Subtracting Fractions**

1) $\frac{9}{9} = 1$
2) $\frac{9}{14}$
3) $\frac{5}{8}$
4) $1\frac{1}{10}$
5) $\frac{17}{20}$
6) $1\frac{1}{4}$
7) $1\frac{1}{5}$
8) $1\frac{1}{15}$
9) $1\frac{8}{21}$
10) $1\frac{1}{3}$
11) $1\frac{7}{30}$
12) $\frac{3}{4}$
13) $\frac{1}{6}$
14) $\frac{5}{8}$
15) $\frac{1}{6}$
16) $\frac{1}{20}$
17) $-\frac{1}{24}$
18) $\frac{3}{28}$
19) $\frac{13}{18}$
20) $\frac{7}{12}$
21) $\frac{19}{24}$
22) $-\frac{1}{15}$
23) $\frac{5}{28}$
24) $\frac{1}{2}$
25) $\frac{3}{28}$
26) $\frac{27}{40}$
27) $\frac{18}{35}$
28) $\frac{5}{16}$
29) $\frac{1}{2}$
30) $\frac{1}{18}$

**Multiplying and Dividing Fractions**

1) $\frac{3}{5}$
2) $\frac{1}{3}$
3) $\frac{1}{20}$
4) $\frac{1}{30}$
5) $\frac{1}{20}$
6) $\frac{1}{9}$
7) $\frac{2}{7}$
8) $\frac{1}{16}$
9) $\frac{3}{8}$
10) $\frac{1}{14}$
11) $\frac{1}{3}$
12) $\frac{1}{6}$
13) 2
14) $\frac{1}{2}$

WWW.MathNotion.Com

# ATI TEAS 6 Subject Test – Mathematics

15) $3\frac{3}{4}$          19) $\frac{26}{49}$          23) $\frac{3}{5}$          27) $\frac{9}{10}$

16) $\frac{2}{5}$           20) $\frac{2}{9}$            24) $\frac{3}{4}$          28) $\frac{8}{15}$

17) $\frac{3}{4}$           21) $\frac{7}{10}$           25) $\frac{5}{18}$         29) $3\frac{3}{4}$

18) $4\frac{1}{2}$          22) 1                        26) $\frac{25}{36}$        30) $\frac{20}{33}$

**Adding and Subtracting Mixed Numbers**

1) $5\frac{1}{2}$           8) $7\frac{9}{10}$           15) $3\frac{1}{4}$         22) $2\frac{1}{16}$

2) 8                         9) $11\frac{29}{35}$         16) $3\frac{13}{15}$       23) $4\frac{1}{24}$

3) $4\frac{1}{2}$           10) $14\frac{19}{48}$         17) $3\frac{1}{8}$         24) $2\frac{71}{72}$

4) $4\frac{7}{12}$          11) $1\frac{1}{2}$           18) $2\frac{13}{15}$       25) $5\frac{33}{35}$

5) $6\frac{5}{12}$          12) $2\frac{1}{5}$           19) $3\frac{7}{30}$        26) $6\frac{32}{63}$

6) $8\frac{13}{15}$         13) $1\frac{2}{9}$           20) $4\frac{3}{35}$

7) $6\frac{16}{21}$         14) $1\frac{9}{14}$          21) $5\frac{1}{12}$

**Multiplying and Dividing Mixed Numbers**

1) $12\frac{3}{8}$          10) 11                       19) $4\frac{4}{27}$

2) $23\frac{1}{9}$          11) $\frac{4}{7}$            20) $2\frac{26}{33}$

3) $35\frac{15}{16}$        12) $1\frac{1}{4}$           21) $3\frac{17}{36}$

4) $8\frac{2}{3}$           13) $4\frac{2}{9}$           22) $\frac{19}{22}$

5) 5                         14) $2\frac{23}{24}$         23) $3\frac{27}{91}$

6) $6\frac{6}{7}$           15) 3                        24) $3\frac{13}{36}$

7) $30\frac{1}{3}$          16) $\frac{11}{45}$          25) $7\frac{1}{5}$

8) $7\frac{6}{7}$           17) $\frac{19}{84}$          26) $4\frac{16}{45}$

9) $21\frac{7}{8}$          18) $\frac{52}{57}$

**Adding and Subtracting Decimals**

1) 10.91                     2) 76.87                     3) 94.95                    4) 17.04

WWW.MathNotion.Com

# ATI TEAS 6 Subject Test – Mathematics

5) 83.93
6) 27.89
7) 50.95
8) 87.17

9) 25.52
10) 2.6
11) 1.78
12) 1.55

13) 3.85
14) 6.6
15) 26.39
16) 6.8

17) 46.58
18) 9.69
19) 52.8

**Multiplying and Dividing Decimals**

1) 0.3
2) 1.32
3) 0.64
4) 0.21
5) 1.11
6) 0.12

7) 7.35
8) 85.1
9) 104.72
10) 50.7
11) 48.3
12) 352.8

13) 0.185
14) 0.746
15) 1.2
16) 24
17) 31
18) 164

19) 0.02215
20) 76.5
21) 1,611
22) 281.5
23) 96.75
24) 931

**Comparing Decimals**

1) >
2) <
3) =
4) >
5) >

6) <
7) =
8) <
9) >
10) >

11) =
12) <
13) <
14) >
15) >

16) =
17) >
18) <
19) >
20) =

**Rounding Decimals**

1) 28
2) 7
3) 16
4) 9
5) 8
6) 53
7) 31.8
8) 14.4

9) 94.7
10) 77.9
11) 13.2
12) 59.9
13) 8.43
14) 23.81
15) 55.38
16) 231.91

17) 62.24
18) 19.45
19) 15.543
20) 34.626
21) 243.865
22) 80.453
23) 67.198
24) 72.368

WWW.MathNotion.Com

ATI TEAS 6 Subject Test – Mathematics

# Chapter 3 : Proportions, Ratios, and Percent

## Topics that you'll practice in this chapter:

- ✓ Simplifying Ratios
- ✓ Proportional Ratios
- ✓ Similarity and Ratios
- ✓ Ratio and Rates Word Problems
- ✓ Percentage Calculations
- ✓ Percent Problems
- ✓ Discount, Tax and Tip
- ✓ Percent of Change
- ✓ Simple Interest

*Without mathematics, there's nothing you can do. Everything around you is mathematics. Everything around you is numbers." – Shakuntala Devi*

ATI TEAS 6 Subject Test – Mathematics

# Simplifying Ratios

✎ Reduce each ratio.

1) $15:20 =$ ___ : ___

2) $7:70 =$ ___ : ___

3) $16:28 =$ ___ : ___

4) $7:21 =$ ___ : ___

5) $4:40 =$ ___ : ___

6) $6:48 =$ ___ : ___

7) $16:64 =$ ___ : ___

8) $10:25 =$ ___ : ___

9) $8:48 =$ ___ : ___

10) $49:63 =$ ___ : ___

11) $18:27 =$ ___ : ___

12) $35:10 =$ ___ : ___

13) $90:9 =$ ___ : ___

14) $24:32 =$ ___ : ___

15) $7:56 =$ ___ : ___

16) $45:63 =$ ___ : ___

17) $56:72 =$ ___ : ___

18) $26:13 =$ ___ : ___

19) $15:45 =$ ___ : ___

20) $28:4 =$ ___ : ___

21) $24:48 =$ ___ : ___

22) $30:24 =$ ___ : ___

23) $70:140 =$ ___ : ___

24) $6:180 =$ ___ : ___

✎ Write each ratio as a fraction in simplest form.

25) $6:12 =$

26) $30:50 =$

27) $15:35 =$

28) $9:27 =$

29) $8:24 =$

30) $18:84 =$

31) $7:14 =$

32) $7:35 =$

33) $40:96 =$

34) $12:54 =$

35) $44:52 =$

36) $12:27 =$

37) $15:180 =$

38) $39:143 =$

39) $20:300 =$

40) $30:120 =$

41) $56:42 =$

42) $26:130 =$

43) $66:123 =$

44) $70:630 =$

45) $75:125 =$

# ATI TEAS 6 Subject Test – Mathematics

## Proportional Ratios

✎ **Fill in the blanks; Calculate each proportion.**

1) $3:8 = \underline{\phantom{xx}} : 48$

2) $2:5 = 20:\underline{\phantom{xx}}$

3) $1:9 = \underline{\phantom{xx}} : 81$

4) $6:7 = 12:\underline{\phantom{xx}}$

5) $9:2 = 63:\underline{\phantom{xx}}$

6) $8:7 = \underline{\phantom{xx}} : 49$

7) $20:3 = \underline{\phantom{xx}} : 15$

8) $1:3 = \underline{\phantom{xx}} : 75$

9) $7:6 = \underline{\phantom{xx}} : 60$

10) $8:5 = \underline{\phantom{xx}} : 45$

11) $3:10 = 60:\underline{\phantom{xx}}$

12) $6:11 = 42:\underline{\phantom{xx}}$

✎ **State if each pair of ratios form a proportion.**

13) $\frac{3}{20}$ and $\frac{9}{60}$

14) $\frac{1}{7}$ and $\frac{6}{42}$

15) $\frac{3}{7}$ and $\frac{24}{56}$

16) $\frac{4}{9}$ and $\frac{12}{18}$

17) $\frac{1}{9}$ and $\frac{12}{81}$

18) $\frac{7}{8}$ and $\frac{21}{28}$

19) $\frac{9}{13}$ and $\frac{27}{39}$

20) $\frac{1}{8}$ and $\frac{8}{64}$

21) $\frac{6}{19}$ and $\frac{30}{85}$

22) $\frac{5}{9}$ and $\frac{40}{81}$

23) $\frac{9}{14}$ and $\frac{108}{168}$

24) $\frac{15}{23}$ and $\frac{360}{552}$

✎ **Calculate each proportion.**

25) $\frac{20}{25} = \frac{32}{x}$, $x = \underline{\phantom{xx}}$

26) $\frac{1}{8} = \frac{32}{x}$, $x = \underline{\phantom{xx}}$

27) $\frac{15}{5} = \frac{21}{x}$, $x = \underline{\phantom{xx}}$

28) $\frac{1}{7} = \frac{x}{294}$, $x = \underline{\phantom{xx}}$

29) $\frac{7}{9} = \frac{x}{81}$, $x = \underline{\phantom{xx}}$

30) $\frac{1}{5} = \frac{13}{x}$, $x = \underline{\phantom{xx}}$

31) $\frac{9}{5} = \frac{36}{x}$, $x = \underline{\phantom{xx}}$

32) $\frac{6}{13} = \frac{48}{x}$, $x = \underline{\phantom{xx}}$

33) $\frac{5}{8} = \frac{x}{88}$, $x = \underline{\phantom{xx}}$

34) $\frac{4}{15} = \frac{x}{240}$, $x = \underline{\phantom{xx}}$

35) $\frac{9}{19} = \frac{x}{266}$, $x = \underline{\phantom{xx}}$

36) $\frac{7}{15} = \frac{x}{270}$, $x = \underline{\phantom{xx}}$

WWW.MathNotion.Com

# ATI TEAS 6 Subject Test – Mathematics

## Similarity and Ratios

✏️ **Each pair of figures is similar. Find the missing side.**

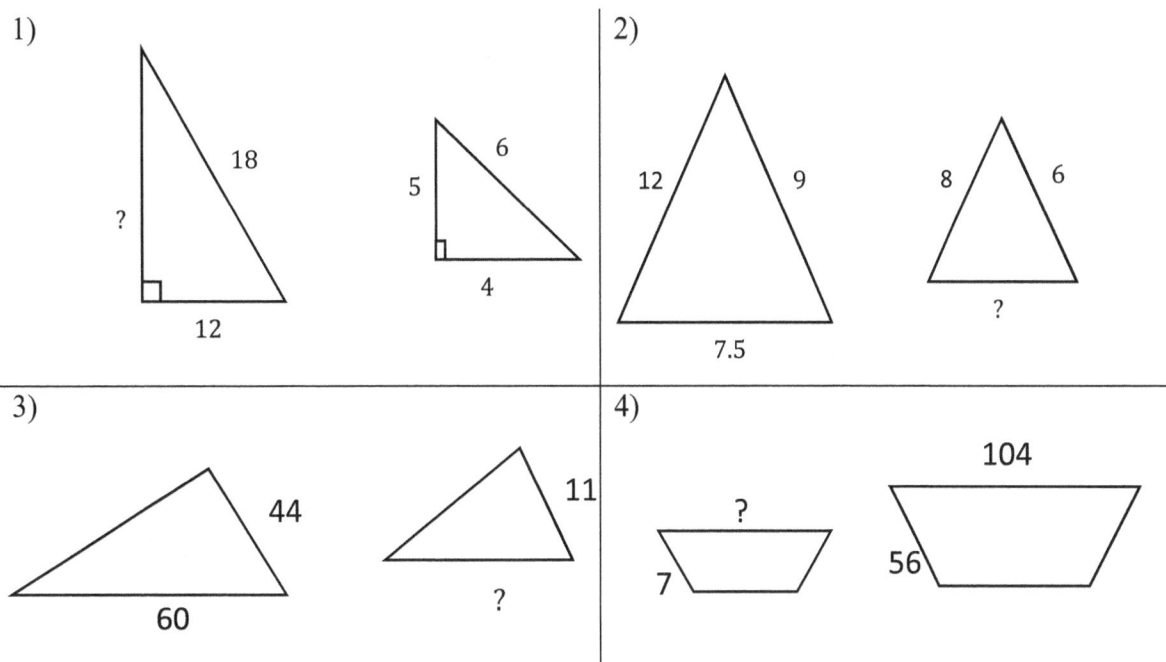

✏️ **Calculate.**

5) Two rectangles are similar. The first is 24 feet wide and 120 feet long. The second is 30 feet wide. What is the length of the second rectangle? _____

6) Two rectangles are similar. One is 5 meters by 36 meters. The longer side of the second rectangle is 90 meters. What is the other side of the second rectangle? _____

7) A building casts a shadow 25 ft long. At the same time a girl 10 ft tall casts a shadow 5 ft long. How tall is the building? _____

8) The scale of a map of Texas is 4 inches: 32 miles. If you measure the distance from Dallas to Martin County as 38.4 inches, approximately how far is Martin County from Dallas? _____

WWW.MathNotion.Com

# ATI TEAS 6 Subject Test – Mathematics

## Ratio and Rates Word Problems

✍ **Find the answer for each word problem.**

1) Mason has 24 red cards and 36 green cards. What is the ratio of Mason's red cards to his green cards? _____

2) In a party, 45 soft drinks are required for every 54 guests. If there are 378 guests, how many soft drinks is required? _____

3) In Mason's class, 42 of the students are tall and 24 are short. In Michael's class 84 students are tall and 48 students are short. Which class has a higher ratio of tall to short students? _____

4) The price of 5 apples at the Quick Market is $4.6. The price of 7 of the same apples at Walmart is $5.95. Which place is the better buy? _____

5) The bakers at a Bakery can make 90 bagels in 3 hours. How many bagels can they bake in 24 hours? What is that rate per hour? _____

6) You can buy 5 cans of green beans at a supermarket for $5.75. How much does it cost to buy 45 cans of green beans? _____

7) The ratio of boys to girls in a class is 4: 7. If there are 32 boys in the class, how many girls are in that class? _____

8) The ratio of red marbles to blue marbles in a bag is 3: 7. If there are 50 marbles in the bag, how many of the marbles are red? _____

WWW.MathNotion.Com

ATI TEAS 6 Subject Test – Mathematics

# Percentage Calculations

✍ **Calculate the given percent of each value.**

1) 3% of 60 = ___

2) 20% of 32 = ___

3) 4% of 72 = ___

4) 16% of 32 = ___

5) 25% of 124 = ___

6) 35% of 56 = ___

7) 15% of 20 = ___

8) 14% of 150 = ___

9) 80% of 50 = ___

10) 12% of 115 = ___

11) 72% of 250 = ___

12) 52% of 500 = ___

13) 70% of 400 = ___

14) 27% of 145 = ___

15) 90% of 64 = ___

16) 60% of 55 = ___

17) 22% of 210 = ___

18) 8% of 235 = ___

✍ **Calculate the percent of each given value.**

19) ___% of 25 = 5

20) ___% of 40 = 20

21) ___% of 25 = 2

22) ___% of 50 = 16

23) ___% of 250 = 5

24) ___% of 40 = 32

25) ___% of 125 = 20

26) ___% of 700 = 49

27) ___% of 350 = 49

28) ___% of 500 = 210

✍ **Calculate each percent problem.**

29) A Cinema has 250 seats. 60 seats were sold for the current movie. What percent of seats are empty? _____ %

30) There are 68 boys and 92 girls in a class. 75% of the students in the class take the bus to school. How many students do not take the bus to school? ____

ATI TEAS 6 Subject Test – Mathematics

# Percent Problems

✎ **Calculate each problem.**

1) 9 is what percent of 45? ___%

2) 60 is what percent of 120? ___%

3) 10 is what percent of 200? ___%

4) 15 is what percent of 125? ___%

5) 10 is what percent of 400? ___%

6) 66 is what percent of 55? ___%

7) 40 is what percent of 160? ___%

8) 40 is what percent of 50? ___%

9) 120 is what percent of 800? ___%

10) 78 is what percent of 120? ___%

11) 36 is what percent of 144? ___%

12) 17 is what percent of 85? ___%

13) 90 is what percent of 900? ___%

14) 36 is what percent of 16? ___%

15) 63 is what percent of 14? ___%

16) 18 is what percent of 60? ___%

17) 126 is what percent of 200? ___%

18) 232 is what percent of 40? ___%

✎ **Calculate each percent word problem.**

19) There are 40 employees in a company. On a certain day, 25 were present. What percent showed up for work? ____%

20) A metal bar weighs 60 ounces. 25% of the bar is gold. How many ounces of gold are in the bar? _____

21) A crew is made up of 12 women; the rest are men. If 15% of the crew are women, how many people are in the crew? _____

22) There are 40 students in a class and 8 of them are girls. What percent are boys? ____%

23) The Royals softball team played 400 games and won 280 of them. What percent of the games did they lose? ____%

ATI TEAS 6 Subject Test – Mathematics

# Discount, Tax and Tip

✎ **Find the selling price of each item.**

1) Original price of a computer: $420
   Tax: 8%  Selling price: $_____

2) Original price of a laptop: $280
   Tax: 4%  Selling price: $_____

3) Original price of a sofa: $820
   Tax: 5%  Selling price: $_____

4) Original price of a car: $15,800
   Tax: 3.6%   Selling price: $_____

5) Original price of a Table: $250
   Tax: 9%  Selling price: $_____

6) Original price of a house: $630,000
   Tax: 1.8%   Selling price: $_____

7) Original price of a tablet: $450
   Discount: 30%   Selling price: $____

8) Original price of a chair: $390
   Discount: 8%   Selling price: $____

9) Original price of a book: $75
   Discount: 42%   Selling price: $____

10) Original price of a cellphone: $820
    Discount: 23%   Selling price: $___

11) Food bill: $45
    Tip: 15%         Price: $_____

12) Food bill: $32
    Tipp: 20%        Price: $_____

13) Food bill: $90
    Tip: 35%         Price: $_____

14) Food bill: $42
    Tipp: 12%    Price: $_____

✎ **Find the answer for each word problem.**

15) Nicolas hired a moving company. The company charged $500 for its services, and Nicolas gives the movers a 40% tip. How much does Nicolas tip the movers? $_____

16) Mason has lunch at a restaurant and the cost of his meal is $90. Mason wants to leave a 25% tip. What is Mason's total bill including tip? $_____

17) The sales tax in Texas is 19.80% and an item costs $350. How much is the tax? $_____

18) The price of a table at Best Buy is $680. If the sales tax is 5%, what is the final price of the table including tax? $_____

WWW.MathNotion.Com

# ATI TEAS 6 Subject Test – Mathematics

## Percent of Change

✒ **Find each percent of change.**

1) From 150 to 450. ___ %

2) From 50 ft to 250 ft. ___ %

3) From $60 to $360. ___ %

4) From 60 cm to 180 cm. ___ %

5) From 15 to 45. ___ %

6) From 80 to 16. ___ %

7) From 120 to 360. ___ %

8) From 900 to 450. ___ %

9) From 1,000 to 200. ___ %

10) From 144 to 36. ___ %

✒ **Calculate each percent of change word problem.**

11) Bob got a raise, and his hourly wage increased from $42 to $63. What is the percent increase? ___ %

12) The price of a pair of shoes increases from $50 to $61. What is the percent increase? ___ %

13) At a coffee shop, the price of a cup of coffee increased from $4.80 to $5.76. What is the percent increase in the cost of the coffee? ___ %

14) 51 cm are cut from 85 cm board. What is the percent decrease in length? ___ %

15) In a class, the number of students has been increased from 54 to 81. What is the percent increase? ___ %

16) The price of gasoline rises from $24.40 to $30.50 in one month. By what percent did the gas price rise? ___ %

17) A shirt was originally priced at $38. It went on sale for $24.70. What was the percent that the shirt was discounted? ___ %

WWW.MathNotion.Com

ATI TEAS 6 Subject Test – Mathematics

# Simple Interest

✍ **Determine the simple interest for these loans.**

1) $480 at 11% for 3 years. $ _____

2) $4,200 at 7% for 4 years. $ _____

3) $2,500 at 20% for 3 years. $ _____

4) $6,800 at 3.9% for 4 months. $ _____

5) $800 at 6% for 7 months. $ _____

6) $36,000 at 4.2% for 6 years. $ _____

7) $6,500 at 7% for 4 years. $ _____

8) $850 at 9.5% for 2 years. $ _____

9) $1,200 at 5.8% for 9 months. $ ___

10) $3,000 at 4.5% for 7 years. $ _____

✍ **Calculate each simple interest word problem.**

11) A new car, valued at $22,000, depreciates at 8.5% per year. What is the value of the car one year after purchase? $_____

12) Sara puts $9,000 into an investment yielding 6% annual simple interest; she left the money in for three years. How much interest does Sara get at the end of those three years? $_____

13) A bank is offering 12% simple interest on a savings account. If you deposit $16,400, how much interest will you earn in two years? $_____

14) $720 interest is earned on a principal of $6,000 at a simple interest rate of 4% interest per year. For how many years was the principal invested? _____

15) In how many years will $2,200 yield an interest of $440 at 4% simple interest? _____

16) Jim invested $8,000 in a bond at a yearly rate of 4.5%. He earned $1,440 in interest. How long was the money invested? _____

WWW.MathNotion.Com

# ATI TEAS 6 Subject Test – Mathematics

## Answers of Worksheets

**Simplifying Ratios**

1) 3 : 4
2) 1 : 10
3) 4 : 7
4) 1 : 3
5) 1 : 10
6) 1 : 8
7) 2 : 8
8) 2 : 5
9) 1 : 6
10) 7 : 9
11) 2 : 3
12) 7 : 2
13) 10 : 1
14) 3 : 4
15) 1 : 8
16) 5 : 7
17) 7 : 9
18) 2 : 1
19) 1 : 3
20) 7 : 1
21) 1 : 2
22) 5 : 4
23) 1 : 2
24) 1 : 30
25) $\frac{1}{2}$
26) $\frac{3}{5}$
27) $\frac{3}{7}$
28) $\frac{1}{3}$
29) $\frac{1}{3}$
30) $\frac{3}{14}$
31) $\frac{1}{2}$
32) $\frac{1}{5}$
33) $\frac{5}{12}$
34) $\frac{2}{9}$
35) $\frac{11}{13}$
36) $\frac{4}{9}$
37) $\frac{1}{12}$
38) $\frac{3}{11}$
39) $\frac{1}{15}$
40) $\frac{1}{4}$
41) $\frac{4}{3}$
42) $\frac{1}{5}$
43) $\frac{22}{41}$
44) $\frac{1}{9}$
45) $\frac{3}{5}$

**Proportional Ratios**

1) 18
2) 50
3) 9
4) 14
5) 14
6) 56
7) 100
8) 25
9) 70
10) 72
11) 200
12) 77
13) Yes
14) Yes
15) Yes
16) No
17) No
18) No
19) Yes
20) Yes
21) No
22) No
23) Yes
24) Yes
25) 40
26) 256
27) 7
28) 42
29) 63
30) 65
31) 20
32) 104
33) 55
34) 64
35) 126
36) 126

**Similarity and ratios**

1) 15
2) 5
3) 15
4) 13
5) 150 feet
6) 12.5 meters
7) 50 feet
8) 307.2 miles

**Ratio and Rates Word Problems**

1) 2 : 3
2) 315

WWW.MathNotion.Com

# ATI TEAS 6 Subject Test – Mathematics

3) The ratio for both classes is 7 to 4.
4) Walmart is a better buy.
5) 720, the rate is 30 per hour.
6) $51.75
7) 56
8) 15

**Percentage Calculations**

1) 1.8
2) 6.4
3) 2.88
4) 5.12
5) 31
6) 19.6
7) 3
8) 21
9) 40
10) 13.8
11) 180
12) 260
13) 280
14) 39.15
15) 57.6
16) 33
17) 46.2
18) 18.8
19) 20%
20) 50%
21) 8%
22) 32%
23) 2%
24) 80%
25) 16%
26) 7%
27) 14%
28) 42%
29) 76%
30) 40

**Percent Problems**

1) 20%
2) 50%
3) 5%
4) 12%
5) 2.5%
6) 120%
7) 25%
8) 80%
9) 15%
10) 65%
11) 25%
12) 20%
13) 10%
14) 225%
15) 450%
16) 30%
17) 63%
18) 580%
19) 62.5%
20) 15 ounces
21) 80
22) 80%
23) 30%

**Discount, Tax and Tip**

1) $453.60
2) $291.20
3) $861.00
4) $16,368.80
5) $272.50
6) $641,340
7) $315.00
8) $358.80
9) $43.50
10) $631.40
11) $51.75
12) $38.40
13) $121.50
14) $47.04
15) $200.00
16) $112.50
17) $69.30
18) $714.00

# ATI TEAS 6 Subject Test – Mathematics

**Percent of Change**

1) 200%
2) 400%
3) 500%
4) 200%
5) 200%
6) 80%
7) 200%
8) 50%
9) 80%
10) 75%
11) 50%
12) 22%
13) 20%
14) 60%
15) 50%
16) 25%
17) 35%

**Simple Interest**

1) $158.40
2) $1,176.00
3) $1,500.00
4) $88.40
5) $28.00
6) $9,072.00
7) $1,820.00
8) $161.50
9) $52.20
10) $945.00
11) $20,130.00
12) $1,620.00
13) $3,936.00
14) 3 years
15) 5 years
16) 4 years

# Chapter 4:
# Exponents and Radicals Expressions

## Topics that you'll practice in this chapter:

- ✓ Multiplication Property of Exponents
- ✓ Zero and Negative Exponents
- ✓ Division Property of Exponents
- ✓ Powers of Products and Quotients
- ✓ Negative Exponents and Negative Bases
- ✓ Scientific Notation
- ✓ Square Roots
- ✓ Simplifying Radical Expressions

*Mathematics is no more computation than typing is literature.*
  *– John Allen Paulos*

ATI TEAS 6 Subject Test – Mathematics

# Multiplication Property of Exponents

✎ Simplify and write the answer in exponential form.

1) $4 \times 4^5 =$

2) $8^4 \times 8 =$

3) $7^3 \times 7^3 =$

4) $9^2 \times 9^2 =$

5) $2^2 \times 2^4 \times 2 =$

6) $5 \times 5^3 \times 5^3 =$

7) $4^3 \times 4^2 \times 4 \times 4 =$

8) $5x \times x =$

9) $x^3 \times x^3 =$

10) $x^7 \times x^2 =$

11) $x^4 \times x^3 \times x^2 =$

12) $10x \times 3x =$

13) $4x^3 \times 4x^3 =$

14) $7x^3 \times x =$

15) $3x^2 \times 4x^2 \times x^2 =$

16) $5x^4 \times x^4 =$

17) $2x^8 \times 2x =$

18) $6x \times x^5 =$

19) $4x^2 \times 6x^6 =$

20) $5yx^3 \times 4x =$

21) $7x^3 \times y^5 x^7 =$

22) $y^2 x^3 \times y^5 x^4 =$

23) $3x^5 \times 4x^3 y^4 =$

24) $4x^4 \times 9x^2 y^5 =$

25) $5x^3 y^4 \times 6x^8 y^2 =$

26) $8x^3 y^6 \times 4xy^3 =$

27) $2xy^5 \times 6x^3 y^3 =$

28) $4x^5 y^2 \times 4x^2 y^8 =$

29) $7x \times 3y^8 x^2 \times y^5 =$

30) $x^3 \times 2y^3 x^4 \times 2y =$

31) $3yx^4 \times 3y^4 x \times 3xy^3 =$

32) $6y^3 \times 2y^2 x^4 \times 10yx^5 =$

WWW.MathNotion.Com

# ATI TEAS 6 Subject Test – Mathematics

## Zero and Negative Exponents

✎ **Evaluate the following expressions.**

1) $1^{-5} =$

2) $4^{-1} =$

3) $0^{10} =$

4) $1^{15} =$

5) $5^{-2} =$

6) $3^{-3} =$

7) $9^{-1} =$

8) $10^{-2} =$

9) $12^{-2} =$

10) $2^{-5} =$

11) $3^{-4} =$

12) $2^{-4} =$

13) $6^{-3} =$

14) $10^{-3} =$

15) $30^{-1} =$

16) $15^{-2} =$

17) $4^{-3} =$

18) $2^{-7} =$

19) $5^{-3} =$

20) $4^{-4} =$

21) $3^{-5} =$

22) $10^{-4} =$

23) $2^{-10} =$

24) $8^{-3} =$

25) $20^{-2} =$

26) $14^{-2} =$

27) $9^{-3} =$

28) $100^{-2} =$

29) $5^{-4} =$

30) $4^{-6} =$

31) $(\frac{1}{4})^{-3} =$

32) $(\frac{1}{6})^{-2} =$

33) $(\frac{1}{7})^{-2} =$

34) $(\frac{2}{3})^{-3} =$

35) $(\frac{1}{13})^{-2} =$

36) $(\frac{7}{12})^{-2} =$

37) $(\frac{1}{6})^{-3} =$

38) $(\frac{1}{300})^{-2} =$

39) $(\frac{2}{9})^{-2} =$

40) $(\frac{7}{5})^{-1} =$

41) $(\frac{13}{23})^{0} =$

42) $(\frac{1}{4})^{-5} =$

# ATI TEAS 6 Subject Test – Mathematics

## Division Property of Exponents

✎ Simplify.

1) $\dfrac{5^6}{5^7} =$

2) $\dfrac{8^8}{8^6} =$

3) $\dfrac{4^5}{4} =$

4) $\dfrac{3}{3^5} =$

5) $\dfrac{x}{x^6} =$

6) $\dfrac{3 \times 3^2}{3^2 \times 3^5} =$

7) $\dfrac{9^4}{9^2} =$

8) $\dfrac{10 \times 10^9}{10^2 \times 10^7} =$

9) $\dfrac{7^5 \times 7^7}{7^4 \times 7^8} =$

10) $\dfrac{15x}{30x^6} =$

11) $\dfrac{3x^9}{4x^4} =$

12) $\dfrac{15x^8}{10x^9} =$

13) $\dfrac{42x^5}{6y^9} =$

14) $\dfrac{36y^8}{4x^4y^5} =$

15) $\dfrac{2x^7}{9x} =$

16) $\dfrac{49x^8y^6}{7x^9} =$

17) $\dfrac{48x^2}{24x^6y^{12}} =$

18) $\dfrac{30yx^5}{6yx^7} =$

19) $\dfrac{19x^7y}{38x^{12}y^4} =$

20) $\dfrac{9x^8}{63x^8} =$

21) $\dfrac{9x^{-9}}{4x^{-3}} =$

WWW.MathNotion.Com

**ATI TEAS 6 Subject Test – Mathematics**

# Powers of Products and Quotients

✎ **Simplify.**

1) $(4^3)^2 =$

2) $(2^3)^4 =$

3) $(2 \times 2^3)^2 =$

4) $(5 \times 5^5)^6 =$

5) $(19^4 \times 19^2)^3 =$

6) $(2^3 \times 2^4)^4 =$

7) $(5 \times 5^2)^2 =$

8) $(4^4)^4 =$

9) $(8x^5)^2 =$

10) $(3x^2 y^4)^4 =$

11) $(7x^5 y^2)^2 =$

12) $(5x^4 y^4)^3 =$

13) $(2x^3 y^3)^5 =$

14) $(10x^3 y^4)^3 =$

15) $(13y^3 y)^2 =$

16) $(5x^6 x^4)^2 =$

17) $(6x^7 y^6)^3 =$

18) $(12x^5 x^7)^2 =$

19) $(2x^4 \times 2x)^4 =$

20) $(2x^4 y^3)^5 =$

21) $(15x^7 y^2)^2 =$

22) $(8x^3 y^5)^3 =$

23) $(3x \times 2y^2)^4 =$

24) $\left(\frac{4x}{x^5}\right)^2 =$

25) $\left(\frac{x^4 y^5}{x^3 y^5}\right)^9 =$

26) $\left(\frac{36xy}{6x^5}\right)^3 =$

27) $\left(\frac{x^7}{x^8 y^2}\right)^6 =$

28) $\left(\frac{xy^4}{x^3 y^6}\right)^{-3} =$

29) $\left(\frac{5xy^8}{x^3}\right)^2 =$

30) $\left(\frac{xy^6}{2xy^3}\right)^{-4} =$

WWW.MathNotion.Com

# ATI TEAS 6 Subject Test – Mathematics

## Negative Exponents and Negative Bases

✎ **Simplify.**

1) $-9^{-1} =$

2) $-9^{-2} =$

3) $-2^{-5} =$

4) $-x^{-7} =$

5) $11x^{-1} =$

6) $-8x^{-3} =$

7) $-12x^{-5} =$

8) $-9x^{-8}y^{-6} =$

9) $32x^{-5}y^{-1} =$

10) $10a^{-9}b^{-3} =$

11) $-17x^4y^{-6} =$

12) $-\dfrac{25}{x^{-5}} =$

13) $-\dfrac{13x}{a^{-7}} =$

14) $\left(-\dfrac{1}{3}\right)^{-4} =$

15) $\left(-\dfrac{3}{4}\right)^{-2} =$

16) $-\dfrac{14}{a^{-6}b^{-3}} =$

17) $-\dfrac{7x}{x^{-8}} =$

18) $-\dfrac{a^{-9}}{b^{-5}} =$

19) $-\dfrac{11}{x^{-5}} =$

20) $\dfrac{8b}{-16c^{-6}} =$

21) $\dfrac{12ab}{a^{-4}b^{-3}} =$

22) $-\dfrac{8n^{-4}}{32p^{-7}} =$

23) $\dfrac{16ab^{-6}}{-6c^{-5}} =$

24) $\left(\dfrac{10a}{5c}\right)^{-4} =$

25) $\left(-\dfrac{12x}{4yz}\right)^{-3} =$

26) $\dfrac{8ab^{-7}}{-5c^{-3}} =$

27) $\left(-\dfrac{x^4}{x^5}\right)^{-5} =$

28) $\left(-\dfrac{x^{-2}}{7x^3}\right)^{-2} =$

29) $\left(-\dfrac{x^{-4}}{x^2}\right)^{-6} =$

# ATI TEAS 6 Subject Test – Mathematics

## Scientific Notation

✏️ **Write each number in scientific notation.**

1) $0.223 =$

2) $0.09 =$

3) $4.5 =$

4) $900 =$

5) $2,000 =$

6) $0.006 =$

7) $33 =$

8) $9,400 =$

9) $1,470 =$

10) $52,000 =$

11) $8,000,000 =$

12) $0.00009 =$

13) $2,158,000 =$

14) $0.0039 =$

15) $0.000075 =$

16) $4,300,000 =$

17) $130,000 =$

18) $4,000,000,000 =$

19) $0.00009 =$

20) $0.0039 =$

✏️ **Write each number in standard notation.**

21) $4 \times 10^{-1} =$

22) $1.2 \times 10^{-3} =$

23) $2.7 \times 10^{5} =$

24) $6 \times 10^{-4} =$

25) $3.6 \times 10^{-3} =$

26) $5.5 \times 10^{5} =$

27) $3.2 \times 10^{4} =$

28) $3.88 \times 10^{6} =$

29) $7 \times 10^{-6} =$

30) $4.2 \times 10^{-7} =$

ATI TEAS 6 Subject Test – Mathematics

# Square Roots

✎ **Find the value each square root.**

1) $\sqrt{16} =$ ___

2) $\sqrt{25} =$ ___

3) $\sqrt{1} =$ ___

4) $\sqrt{64} =$ ___

5) $\sqrt{0} =$ ___

6) $\sqrt{196} =$ ___

7) $\sqrt{4} =$ ___

8) $\sqrt{256} =$ ___

9) $\sqrt{36} =$ ___

10) $\sqrt{289} =$ ___

11) $\sqrt{169} =$ ___

12) $\sqrt{144} =$ ___

13) $\sqrt{100} =$ ___

14) $\sqrt{1,600} =$ ___

15) $\sqrt{2,500} =$ ___

16) $\sqrt{324} =$ ___

17) $\sqrt{529} =$ ___

18) $\sqrt{20} =$ ___

19) $\sqrt{625} =$ ___

20) $\sqrt{18} =$ ___

21) $\sqrt{50} =$ ___

22) $\sqrt{1,024} =$ ___

23) $\sqrt{160} =$ ___

24) $\sqrt{32} =$ ___

✎ **Evaluate.**

25) $\sqrt{4} \times \sqrt{25} =$ _____

26) $\sqrt{36} \times \sqrt{49} =$ _____

27) $\sqrt{6} \times \sqrt{6} =$ _____

28) $\sqrt{13} \times \sqrt{13} =$ _____

29) $2\sqrt{5} \times 3\sqrt{5} =$ _____

30) $\sqrt{12} \times \sqrt{3} =$ _____

31) $\sqrt{13} + \sqrt{13} =$ _____

32) $\sqrt{10} + 2\sqrt{10} =$ _____

33) $12\sqrt{7} - 10\sqrt{7} =$ _____

34) $4\sqrt{10} \times 2\sqrt{10} =$ _____

35) $5\sqrt{3} \times 8\sqrt{3} =$ _____

36) $6\sqrt{3} - \sqrt{12} =$ _____

WWW.MathNotion.Com

ATI TEAS 6 Subject Test – Mathematics

# Simplifying Radical Expressions

✎ **Simplify.**

1) $\sqrt{13x^2} =$

2) $\sqrt{75x^2} =$

3) $\sqrt[3]{27a} =$

4) $\sqrt{64x^5} =$

5) $\sqrt{216a} =$

6) $\sqrt[3]{63w^3} =$

7) $\sqrt{192x} =$

8) $\sqrt{125v} =$

9) $\sqrt[3]{128x^2} =$

10) $\sqrt{100x^9} =$

11) $\sqrt{16x^4} =$

12) $\sqrt[3]{500a^5} =$

13) $\sqrt{242} =$

14) $\sqrt{392p^3} =$

15) $\sqrt{8m^6} =$

16) $\sqrt{198x^3y^3} =$

17) $\sqrt{121x^5y^5} =$

18) $\sqrt{16a^6b^3} =$

19) $\sqrt{90x^5y^7} =$

20) $\sqrt[3]{64y^2x^6} =$

21) $10\sqrt{16x^4} =$

22) $6\sqrt{81x^2} =$

23) $\sqrt[3]{56x^2y^6} =$

24) $\sqrt[3]{1,000x^5y^7} =$

25) $8\sqrt{50a} =$

26) $\sqrt[4]{625x^8y} =$

27) $\sqrt{24x^4y^5r^3} =$

28) $5\sqrt{36x^4y^5z^8} =$

29) $3\sqrt[3]{343x^9y^7} =$

30) $5\sqrt{81a^5b^2c^9} =$

31) $\sqrt[4]{625x^8y^{16}} =$

WWW.MathNotion.Com

**ATI TEAS 6 Subject Test – Mathematics**

# Answers of Worksheets

**Multiplication Property of Exponents**

1) $4^6$
2) $8^5$
3) $7^6$
4) $9^4$
5) $2^7$
6) $5^7$
7) $4^7$
8) $5x^2$
9) $x^6$
10) $x^9$
11) $x^9$
12) $30x^2$
13) $16x^6$
14) $7x^4$
15) $12x^6$
16) $5x^8$
17) $4x^9$
18) $6x^6$
19) $24x^8$
20) $20x^4y$
21) $7x^{10}y^5$
22) $x^7y^7$
23) $12x^8y^4$
24) $36x^6y^5$
25) $30x^{11}y^6$
26) $32x^4y^9$
27) $12x^4y^8$
28) $16x^7y^{10}$
29) $21x^3y^{13}$
30) $4x^7y^4$
31) $27x^6y^8$
32) $120x^9y^6$

**Zero and Negative Exponents**

1) $1$
2) $\frac{1}{4}$
3) $0$
4) $1$
5) $\frac{1}{25}$
6) $\frac{1}{27}$
7) $\frac{1}{9}$
8) $\frac{1}{100}$
9) $\frac{1}{144}$
10) $\frac{1}{32}$
11) $\frac{1}{81}$
12) $\frac{1}{16}$
13) $\frac{1}{216}$
14) $\frac{1}{1,000}$
15) $\frac{1}{30}$
16) $\frac{1}{225}$
17) $\frac{1}{64}$
18) $\frac{1}{128}$
19) $\frac{1}{125}$
20) $\frac{1}{256}$
21) $\frac{1}{243}$
22) $\frac{1}{10,000}$
23) $\frac{1}{1,024}$
24) $\frac{1}{512}$
25) $\frac{1}{400}$
26) $\frac{1}{196}$
27) $\frac{1}{729}$
28) $\frac{1}{10,000}$
29) $\frac{1}{625}$
30) $\frac{1}{4,096}$
31) $64$
32) $36$
33) $49$
34) $\frac{27}{8}$
35) $169$
36) $\frac{144}{49}$
37) $216$
38) $90,000$
39) $\frac{81}{4}$
40) $\frac{5}{7}$
41) $1$
42) $1,024$

**Division Property of Exponents**

1) $\frac{1}{5}$
2) $8^2$
3) $4^4$
4) $\frac{1}{3^4}$
5) $\frac{1}{x^5}$
6) $\frac{1}{3^4}$
7) $9^2$
8) $10$
9) $1$
10) $\frac{1}{2x^5}$
11) $\frac{3x^5}{4}$
12) $\frac{3}{2x}$
13) $\frac{7x^5}{y^9}$
14) $\frac{9y^3}{x^4}$

# ATI TEAS 6 Subject Test – Mathematics

15) $\frac{2x^6}{9}$

16) $\frac{7y^6}{x}$

17) $\frac{2}{x^4 y^{12}}$

18) $\frac{5}{x^2}$

19) $\frac{1}{2x^5 y^3}$

20) $\frac{1}{7}$

21) $\frac{9}{4x^6}$

## Powers of Products and Quotients

1) $4^6$
2) $2^{12}$
3) $2^8$
4) $5^{36}$
5) $19^{18}$
6) $2^{28}$
7) $5^6$
8) $4^{16}$
9) $64x^{10}$
10) $81x^8 y^{16}$
11) $49x^{10} y^4$
12) $125x^{12} y^{12}$
13) $32x^{15} y^{15}$
14) $1{,}000 x^9 y^{12}$
15) $169 y^8$
16) $25 x^{20}$
17) $216 x^{21} y^{18}$
18) $144 x^{24}$
19) $256 x^{20}$
20) $32 x^{20} y^{15}$
21) $225 x^{14} y^4$
22) $512 x^9 y^{15}$
23) $1{,}296 x^4 y^8$
24) $\frac{16}{x^8}$
25) $x^9$
26) $\frac{216 y^3}{x^{12}}$
27) $\frac{1}{x^6 y^{12}}$
28) $x^6 y^6$
29) $\frac{25 y^{16}}{x^4}$
30) $\frac{16}{y^{12}}$

## Negative Exponents and Negative Bases

1) $-\frac{1}{9}$
2) $-\frac{1}{81}$
3) $-\frac{1}{32}$
4) $-\frac{1}{x^7}$
5) $\frac{11}{x}$
6) $-\frac{8}{x^3}$
7) $-\frac{12}{x^5}$
8) $-\frac{9}{x^8 y^6}$
9) $\frac{32}{x^5 y}$
10) $\frac{10}{a^9 b^3}$
11) $-\frac{17 x^4}{y^6}$
12) $-25 x^5$
13) $-13 x a^7$
14) $81$
15) $\frac{16}{9}$
16) $-14 a^6 b^3$
17) $-7 x^9$
18) $-\frac{b^5}{a^9}$
19) $-11 x^5$
20) $-\frac{bc^6}{2}$
21) $12 a^5 b^4$
22) $-\frac{p^7}{4n^4}$
23) $-\frac{8ac^5}{3b^6}$
24) $\frac{c^4}{16 a^4}$
25) $\frac{y^3 z^3}{27 x^3}$
26) $-\frac{8ac^3}{5 b^7}$
27) $-x^5$
28) $49 x^{10}$
29) $x^{36}$

## Scientific Notation

1) $2.23 \times 10^{-1}$
2) $9 \times 10^{-2}$
3) $4.5 \times 10^0$
4) $9 \times 10^2$
5) $2 \times 10^3$
6) $6 \times 10^{-3}$
7) $3.3 \times 10^1$
8) $9.4 \times 10^3$
9) $1.47 \times 10^3$

# ATI TEAS 6 Subject Test – Mathematics

10) $5.2 \times 10^4$
11) $8 \times 10^6$
12) $9 \times 10^{-5}$
13) $2.158 \times 10^6$
14) $3.9 \times 10^{-3}$
15) $7.5 \times 10^{-5}$
16) $4.3 \times 10^6$

17) $1.3 \times 10^5$
18) $4 \times 10^9$
19) $9 \times 10^{-5}$
20) $3.9 \times 10^{-3}$
21) $0.4$
22) $0.0012$
23) $270,000$

24) $0.0006$
25) $0.0036$
26) $550,000$
27) $32,000$
28) $3,880,000$
29) $0.000007$
30) $0.00000042$

**Square Roots**

1) 4
2) 5
3) 1
4) 8
5) 0
6) 14
7) 2
8) 16
9) 6

10) 17
11) 13
12) 12
13) 10
14) 40
15) 50
16) 18
17) 23
18) $2\sqrt{5}$

19) 25
20) $3\sqrt{2}$
21) $5\sqrt{2}$
22) 32
23) $4\sqrt{10}$
24) $4\sqrt{2}$
25) 10
26) 42
27) 6

28) 13
29) 30
30) 6
31) $2\sqrt{13}$
32) $3\sqrt{10}$
33) $2\sqrt{7}$
34) 80
35) 120
36) $4\sqrt{3}$

**Simplifying radical expressions**

1) $x\sqrt{13}$
2) $5x\sqrt{3}$
3) $3\sqrt[3]{a}$
4) $8x^2\sqrt{x}$
5) $6\sqrt{6a}$
6) $w\sqrt[3]{63}$
7) $8\sqrt{3x}$
8) $5\sqrt{5v}$
9) $4\sqrt[3]{2x^2}$
10) $10x^4\sqrt{x}$
11) $4x^2$

12) $5a\sqrt[3]{4a^2}$
13) $11\sqrt{2}$
14) $14p\sqrt{2p}$
15) $2m^3\sqrt{2}$
16) $3x.y\sqrt{22xy}$
17) $11x^2y^2\sqrt{xy}$
18) $4a^3b\sqrt{b}$
19) $3x^2y^3\sqrt{10xy}$
20) $4x^2\sqrt[3]{y^2}$
21) $40x^2$
22) $54x$

23) $2y^2\sqrt[3]{7x^2}$
24) $10xy^2\sqrt[3]{x^2y}$
25) $40\sqrt{2a}$
26) $5x^2\sqrt[4]{y}$
27) $2x^2y^2r\sqrt{6yr}$
28) $30x^2y^2z^4\sqrt{y}$
29) $21x^3y^2\sqrt[3]{y}$
30) $45a^2bc^4\sqrt{ac}$
31) $5x^2y^4$

ATI TEAS 6 Subject Test – Mathematics

# Chapter 5:
# Algebraic Expressions

## Topics that you'll practice in this chapter:

- ✓ Simplifying Variable Expressions
- ✓ Simplifying Polynomial Expressions
- ✓ Translate Phrases into an Algebraic Statement
- ✓ The Distributive Property
- ✓ Evaluating One Variable Expressions
- ✓ Evaluating Two Variables Expressions
- ✓ Combining like Terms

*Mathematics is, as it were, a sensuous logic, and relates to philosophy as do the arts, music, and plastic art to poetry.* — *K. Shegel*

ATI TEAS 6 Subject Test – Mathematics

## Simplifying Variable Expressions

✎ **Simplify each expression.**

1) $3(x + 5) =$

2) $(-4)(7x - 5) =$

3) $11x + 5 - 6x =$

4) $-4 - 2x^2 - 6x^2 =$

5) $7 + 13x^2 + 3 =$

6) $3x^2 + 7x + 15x^2 =$

7) $3x^2 - 12x^2 + 4x =$

8) $4x^2 - 8x - 2x =$

9) $6x + 7(3 - 4x) =$

10) $8x + 4(15x - 3) =$

11) $6(-3x - 9) - 17 =$

12) $-11x^2 - (-5x) =$

13) $2x + 7 + 5 - 8x =$

14) $7 + 6x - 11 - 5x =$

15) $27x + 8 - 13 - 5x =$

16) $(-11)(-5x + 2) - 41x =$

17) $19x - 4(4 - 2x) =$

18) $16x + 3(3x + 6) + 10 =$

19) $5(-2x - 4) - 13x =$

20) $16x - 3x(x + 10) =$

21) $17x + 5x(2 - 4x) =$

22) $5x(-4x - 7) + 20x =$

23) $25x - 19 + 4x^2 =$

24) $6x(x - 11) + 25 =$

25) $4x - 5 + 15x + 3x^2 =$

26) $-7x^2 - 11x - 9x =$

27) $10x - 9x^2 - 3x^2 - 7 =$

28) $13 + 3x^2 - 9x^2 - 21x =$

29) $22x + 10x^2 - 15x + 17 =$

30) $4x^2 + 25x + 21x^2 =$

31) $29 - 12x^2 - 23x - 4x^2 =$

32) $22x - 19x - 9x^2 + 30 =$

WWW.MathNotion.Com

ATI TEAS 6 Subject Test – Mathematics

## Simplifying Polynomial Expressions

✏️ **Simplify each polynomial.**

1) $(2x^3 + 8x^2) - (11x + 3x^2) =$ _____

2) $(2x^5 + 7x^3) - (5x^3 + 11x^2) =$ _____

3) $(41x^4 + 5x^2) - (4x^2 + 20x^4) =$ _____

4) $13x - 8x^2 + 4(4x^2 + 3x^3) =$ _____

5) $(4x^3 - 22) + 5(3x^2 - 6x^3) =$ _____

6) $(4x^3 - 3x) - 5(2x^3 + x^4) =$ _____

7) $5(5x - 2x^3) - 2(8x^3 + 5x^2) =$ _____

8) $(3x^2 - 10x) - (5x^3 + 14x^2) =$ _____

9) $5x^3 - (3x^4 + 5x) + 2x^2 =$ _____

10) $11x^4 - (3x^2 + 5x) + 7x =$ _____

11) $(6x^2 - 3x^4) - (10x^4 + 3x^2) =$ _____

12) $2x^2 - 7x^3 + 19x^4 - 22x^3 =$ _____

13) $10x^2 - x^4 + 4x^4 - 32x^3 =$ _____

14) $-5x^2 + 17x^3 - 8x^2 - 6x =$ _____

15) $x^4 - 11x^5 - 30x^4 + 5x^2 =$ _____

16) $21x^3 + 13x - 5x^2 - 11x^3 =$ _____

WWW.MathNotion.Com

ATI TEAS 6 Subject Test – Mathematics

# Translate Phrases into an Algebraic Statement

✎ Write an algebraic expression for each phrase.

1) 9 multiplied by $x$. _____

2) Subtract 11 from $y$. _____

3) 19 divided by $x$. _____

4) 38 decreased by $y$. _____

5) Add $y$ to 40. _____

6) The square of 6. _____

7) $x$ raised to the fifth power. _____

8) The sum of six and a number. _____

9) The difference between fifty-seven and $y$. _____

10) The quotient of nine and a number. _____

11) The quotient of the square of $x$ and 25. _____

12) The difference between $x$ and 6 is 19. _____

13) 10 times $a$ reduced by the square of $b$. _____

14) Subtract the product of $a$ and $b$ from 41. _____

WWW.MathNotion.Com

ATI TEAS 6 Subject Test – Mathematics

# The Distributive Property

✎ Use the distributive property to simply each expression.

1) $4(1 + 2x) =$

2) $2(4 + 7x) =$

3) $3(4x - 4) =$

4) $(2x - 5)(-6) =$

5) $(-3)(x + 6) =$

6) $(4 + 3x)2 =$

7) $(-5)(8 - 3x) =$

8) $-(-5 - 7x) =$

9) $(-6x + 3)(-3) =$

10) $(-4)(x - 7) =$

11) $-(5 - 3x) =$

12) $3(9 + 4x) =$

13) $6(4 + 3x) =$

14) $(-5x + 3)2 =$

15) $(5 - 8x)(-3) =$

16) $(-12)(3x + 3) =$

17) $(5 - 3x)6 =$

18) $4(2 + 6x) =$

19) $8(7x - 3) =$

20) $(-2x + 3)4 =$

21) $(7 - 5x)(-9) =$

22) $(-10)(x - 8) =$

23) $(11 - 4x)3 =$

24) $(-6)(10x - 4) =$

25) $(3 - 9x)(-7) =$

26) $(-9)(x + 9) =$

27) $(-3 + 5x)(-7) =$

28) $(-5)(8 - 10x) =$

29) $12(4x - 8) =$

30) $(-10x + 13)(-3) =$

31) $(-8)(3x - 2) + 4(x + 5) =$

32) $(-8)(x + 4) - (6 + 5x) =$

WWW.MathNotion.Com

ATI TEAS 6 Subject Test – Mathematics

# Evaluating One Variable Expressions

✏️ **Evaluate each expression using the value given.**

1) $8 - x, x = 5$

2) $x - 9, x = 5$

3) $5x + 4, x = 3$

4) $x - 13, x = -4$

5) $12 - x, x = 4$

6) $x + 2, x = 6$

7) $4x + 8, x = 3$

8) $x + (-7), x = -8$

9) $4x + 5, x = 2$

10) $3x + 9, x = -2$

11) $15 + 3x - 7, x = 2$

12) $17 - 3x, x = 3$

13) $8x - 9, x = 4$

14) $5x + 4, x = -3$

15) $10x + 5, x = 3$

16) $14 - 4x, x = -6$

17) $3(5x + 3), x = 9$

18) $4(-3x - 6), x = 3$

19) $7x - 2x + 12, x = 4$

20) $(5x + 6) \div 2, x = 8$

21) $(x + 18) \div 10, x = 12$

22) $5x - 12 + 3x, x = -3$

23) $(6 - 4x)(-3), x = -4$

24) $9x^2 + 3x - 6, x = 2$

25) $x^2 - 10x, x = -5$

26) $3x(7 - 2x), x = 2$

27) $12x + 6 - 2x^2, x = -4$

28) $(-3)(4x - 8 + 3x), x = 3$

29) $(-6) + \frac{x}{4} + 3x, x = 16$

30) $(-6) + \frac{x}{5}, x = 35$

31) $\left(-\frac{45}{x}\right) - 7 + 2x, x = 9$

32) $\left(-\frac{21}{x}\right) - 12 + 4x, x = 7$

# ATI TEAS 6 Subject Test – Mathematics

## Evaluating Two Variables Expressions

✎ **Evaluate each expression using the values given.**

1) $2x - 4y$,

   $x = 4, y = 1$

2) $3x + 5y$,

   $x = -2, y = 2$

3) $-7a + 4b$,

   $a = 2, b = 4$

4) $3x + 5 - y$,

   $x = 5, y = 6$

5) $3z + 12 - 2k$,

   $z = 5, k = 6$

6) $6(-x - 3y)$,

   $x = 5, y = -2$

7) $5a + 3b$,

   $a = 3, b = 4$

8) $7x \div 3y$,

   $x = 3, y = 7$

9) $2x + 15 + 5y$,

   $x = -3, y = 1$

10) $5a - (18 - b)$,

    $a = 2, b = 8$

11) $2z + 20 + 5k$,

    $z = -6, k = 5$

12) $xy + 10 + 4x$,

    $x = 3, y = 5$

13) $2x + 4y - 8 + 5$,

    $x = 5, y = 2$

14) $\left(-\frac{24}{x}\right) + 3 + 2y$,

    $x = 4, y = 6$

15) $(-3)(-3a - 3b)$,

    $a = 4, b = 5$

16) $12 + 4x - 7 - y$,

    $x = 3, y = 5$

17) $11x + 5 - 8y + 6$,

    $x = 5, y = 2$

18) $10 + 2(-4x - 5y)$,

    $x = 5, y = 4$

19) $5x + 13 + 6y$,

    $x = 5, y = 6$

20) $10a - (7a + 3b) - 11$,

    $a = 3, b = 8$

WWW.MathNotion.Com

ATI TEAS 6 Subject Test – Mathematics

## Combining like Terms

✎ **Simplify each expression.**

1) $11x + 3x + 6 =$

2) $8(2x - 6) =$

3) $18x - 7x + 11 =$

4) $(-4)(6x - 7) =$

5) $22x - 10x - 5 =$

6) $32x - 13 + 8x =$

7) $15 - (8x - 11) =$

8) $-24x + 17 - 11x =$

9) $12x - 8 - 6x + 9 =$

10) $21x + 5 - 36 + 12x =$

11) $28x + 3x - 11 =$

12) $(-3x + 4)5 =$

13) $2 + 4x + 9x - 8 =$

14) $6(2x - 5x) - 4 =$

15) $4(5x + 11) + 3x =$

16) $x - 14 - 11x =$

17) $5(10 + 9x) - 8x =$

18) $42x + 17 - 23x =$

19) $(-7x) + 19 + 20x =$

20) $(-7x) - 33 + 29x =$

21) $4(5x + 3) - 19x =$

22) $5(6 - 2x) - 15x =$

23) $-24x + (11 - 18x) =$

24) $(-9) - (6)(7x + 3) =$

25) $(-1)(8x - 10) - 21x =$

26) $-36x + 14 + 27x - 5x =$

27) $3(-13x + 6) - 17x =$

28) $-5x - 42 + 32x =$

29) $37x - 19x + 15 - 9x =$

30) $3(5x + 7x) - 31 =$

31) $14 - 6x - 15 - 9x =$

32) $-2(-5x - 7x) + 27x =$

# ATI TEAS 6 Subject Test – Mathematics

## Answers of Worksheets

**Simplifying Variable Expressions**

1) $3x + 15$
2) $-28x + 20$
3) $5x + 5$
4) $-8x^2 - 4$
5) $13x^2 + 10$
6) $18x^2 + 7x$
7) $-9x^2 + 4x$
8) $4x^2 - 10x$
9) $-22x + 21$
10) $68x - 12$
11) $-18x - 71$
12) $-11x^2 + 5x$
13) $-6x + 12$
14) $x - 4$
15) $22x - 5$
16) $14x - 22$
17) $27x - 16$
18) $25x + 28$
19) $-23x - 20$
20) $-3x^2 - 14x$
21) $-20x^2 + 27x$
22) $-20x^2 - 15x$
23) $4x^2 + 25x - 19$
24) $6x^2 - 66x + 25$
25) $3x^2 + 19x - 5$
26) $-7x^2 - 20x$
27) $-12x^2 + 10x - 7$
28) $-6x^2 - 21x + 13$
29) $10x^2 + 7x + 17$
30) $25x^2 + 25x$
31) $-16x^2 - 23x + 29$
32) $-9x^2 + 3x + 30$

**Simplifying Polynomial Expressions**

1) $2x^3 + 5x^2 - 11x$
2) $2x^5 + 2x^3 - 11x^2$
3) $21x^4 + x^2$
4) $12x^3 + 8x^2 + 13x$
5) $-26x^3 + 15x^2 - 22$
6) $-5x^4 - 6x^3 - 3x$
7) $-26x^3 - 10x^2 + 25x$
8) $-5x^3 - 11x^2 - 10x$
9) $-3x^4 + 5x^3 + 2x^2 - 5x$
10) $11x^4 - 3x^2 + 2x$
11) $-13x^4 + 3x^2$
12) $19x^4 - 29x^3 + 2x^2$
13) $3x^4 - 32x^3 + 10x^2$
14) $17x^3 - 13x^2 - 6x$
15) $-11x^5 - 29x^4 + 5x^2$
16) $10x^3 - 5x^2 + 13x$

**Translate Phrases into an Algebraic Statement**

1) $9x$
2) $y - 11$
3) $\frac{19}{x}$
4) $38 - y$
5) $y + 40$
6) $6^2$
7) $x^5$
8) $6 + x$
9) $57 - y$
10) $\frac{9}{x}$
11) $\frac{x^2}{25}$
12) $x - 6 = 19$
13) $10a - b^2$
14) $41 - ab$

**The Distributive Property**

1) $8x + 4$
2) $14x + 8$
3) $12x - 12$
4) $-12x + 30$
5) $-3x - 18$
6) $6x + 8$
7) $15x - 40$
8) $7x + 5$
9) $18x - 9$
10) $-4x + 28$
11) $3x - 5$
12) $12x + 27$

# ATI TEAS 6 Subject Test – Mathematics

13) $18x + 24$
14) $-10x + 6$
15) $24x - 15$
16) $-36x - 36$
17) $-18x + 30$

18) $24x + 8$
19) $56x - 24$
20) $-8x + 12$
21) $45x - 63$
22) $-10x + 80$

23) $-12x + 33$
24) $-60x + 24$
25) $63x - 21$
26) $-9x - 81$
27) $-35x + 21$

28) $50x - 40$
29) $48x - 96$
30) $30x - 39$
31) $-20x + 36$
32) $-13x - 38$

**Evaluating One Variables**

1) 3
2) $-4$
3) 19
4) $-17$
5) 8
6) 8
7) 20
8) $-15$

9) 13
10) 3
11) 14
12) 8
13) 23
14) $-11$
15) 35
16) 38

17) 144
18) $-60$
19) 32
20) 23
21) 3
22) $-36$
23) $-66$
24) 36

25) 75
26) 18
27) $-74$
28) $-39$
29) 46
30) 1
31) 6
32) 13

**Evaluating Two Variables**

1) 4
2) 4
3) 2
4) 14
5) 15

6) 6
7) 27
8) 1
9) 14
10) 0

11) 33
12) 37
13) 15
14) 9
15) 81

16) 12
17) 50
18) $-70$
19) 74
20) $-26$

**Combining like Terms**

1) $14x + 6$
2) $16x - 48$
3) $11x + 11$
4) $-24x + 28$
5) $12x - 5$
6) $40x - 13$
7) $-8x + 26$
8) $-35x + 17$

9) $6x + 1$
10) $33x - 31$
11) $31x - 11$
12) $-15x + 20$
13) $13x - 6$
14) $-18x - 4$
15) $23x + 44$
16) $-10x - 14$

17) $37x + 50$
18) $19x + 17$
19) $13x + 19$
20) $22x - 33$
21) $x + 12$
22) $-25x + 30$
23) $-42x + 11$
24) $-42x - 27$

25) $-29x + 10$
26) $-14x + 14$
27) $-56x + 18$
28) $27x - 42$
29) $9x + 15$
30) $36x - 31$
31) $-15x - 1$
32) $51x$

ATI TEAS 6 Subject Test – Mathematics

# Chapter 6:
# Equations and Inequalities

**Topics that you'll practice in this chapter:**

- ✓ One–Step Equations
- ✓ Multi–Step Equations
- ✓ Graphing Single–Variable Inequalities
- ✓ One–Step Inequalities
- ✓ Multi-Step Inequalities
- ✓ Systems of Equations
- ✓ Systems of Equations Word Problems

*"Life is a math equation. In order to gain the most, you have to know how to convert negatives into positives." – Anonymous*

ATI TEAS 6 Subject Test – Mathematics

## One–Step Equations

✏️ **Find the answer for each equation.**

1) $3x = 90, x =$ ___

2) $5x = 35, x =$ ___

3) $6x = 24, x =$ ___

4) $24x = 144, x =$ ___

5) $x + 15 = 20, x =$ ___

6) $x - 7 = 4, x =$ ___

7) $x - 9 = 2, x =$ ___

8) $x + 15 = 23, x =$ ___

9) $x - 4 = 13, x =$ ___

10) $12 = 16 + x, x =$ ___

11) $x - 10 = 2, x =$ ___

12) $5 - x = -11, x =$ ___

13) $28 = -6 + x, x =$ ___

14) $x - 20 = -35, x =$ ___

15) $x + 14 = -4, x =$ ___

16) $14 = 28 - x, x =$ ___

17) $7 + x = -7, x =$ ___

18) $x - 16 = 4, x =$ ___

19) $30 = x - 15, x =$ ___

20) $x - 5 = -18, x =$ ___

21) $x - 10 = 24, x =$ ___

22) $x - 20 = -25, x =$ ___

23) $x - 17 = 30, x =$ ___

24) $-70 = x - 28, x =$ ___

25) $x - 9 = 13, x =$ ___

26) $36 = 4x, x =$ ___

27) $x - 35 = 25, x =$ ___

28) $x - 25 = 10, x =$ ___

29) $70 - x = 16, x =$ ___

30) $x - 10 = 14, x =$ ___

31) $17 - x = -13, x =$ ___

32) $x - 9 = -30, x =$ ___

WWW.MathNotion.Com

ATI TEAS 6 Subject Test – Mathematics

## Multi–Step Equations

✏️ **Find the answer for each equation.**

1) $3x + 3 = 9$

2) $-x + 5 = 12$

3) $4x - 8 = 8$

4) $-(3 - x) = 5$

5) $4x - 8 = 16$

6) $12x - 15 = 9$

7) $2x - 18 = 2$

8) $4x + 8 = 16$

9) $24x + 27 = 75$

10) $-14(3 + x) = 14$

11) $-3(2 + x) = 6$

12) $12 = -(x - 7)$

13) $3(3 - x) = 30$

14) $-15 = -(3x + 6)$

15) $40(3 + x) = 40$

16) $5(x - 10) = 25$

17) $-18 = x + 8x$

18) $3x + 25 = -2x - 10$

19) $7(6 + 3x) = -63$

20) $18 - 3x = -4 - 5x$

21) $4 - 6x = 36 + 2x$

22) $15 + 15x = -5 + 5x$

23) $42 = (-6x) - 7 + 7$

24) $21 = 3x - 21 + 4x$

25) $-18 = -6x - 9 + 3x$

26) $5x - 15 = -29 + 6x$

27) $7x - 18 = 4x + 3$

28) $-7 - 4x = 5(4 - x)$

29) $x - 5 = -5(-3 - x)$

30) $13x - 68 = 15x - 102$

31) $-5x - 3 = -3(9 + 3x)$

32) $-2x - 15 = 6x + 17$

WWW.MathNotion.Com

ATI TEAS 6 Subject Test – Mathematics

## Graphing Single–Variable Inequalities

✎ Draw a graph for each inequality.

1) $x > -1$

2) $x \leq 2$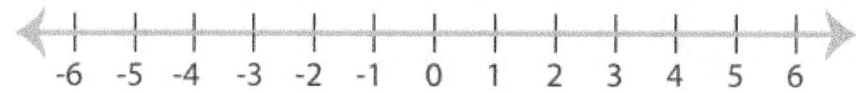

3) $x \geq 0$

4) $x < -3$

5) $x < \frac{1}{2}$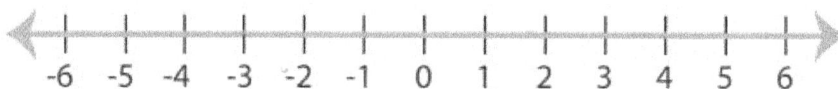

6) $x \leq -2$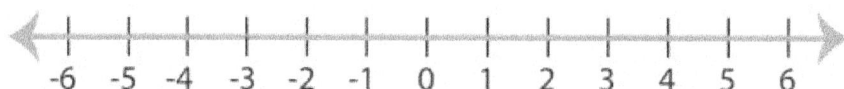

7) $x \leq 3$

8) $x \geq -\frac{7}{2}$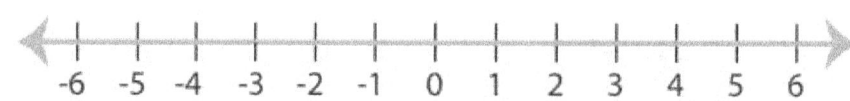

WWW.MathNotion.Com

ATI TEAS 6 Subject Test – Mathematics

# One–Step Inequalities

✎ Find the answer for each inequality and graph it.

1) $x + 4 \geq 4$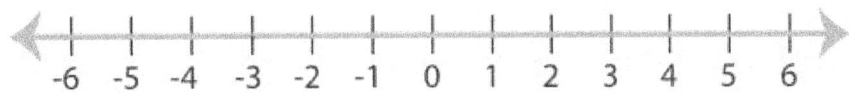

2) $x - 5 \leq 2$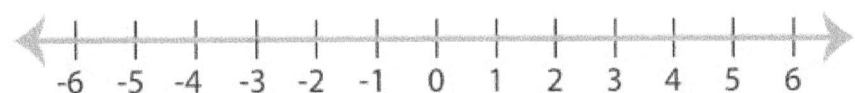

3) $5x > 35$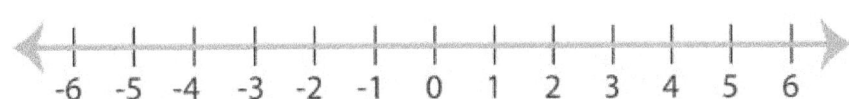

4) $9 + x \leq 11$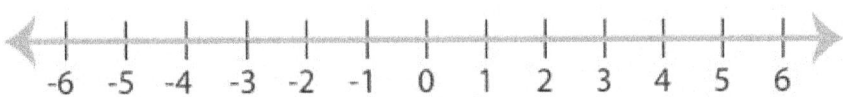

5) $x - 5 < -9$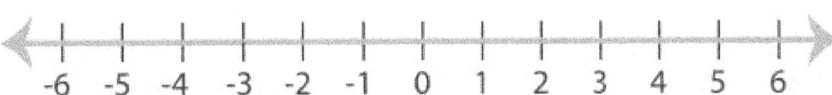

6) $9x \geq 72$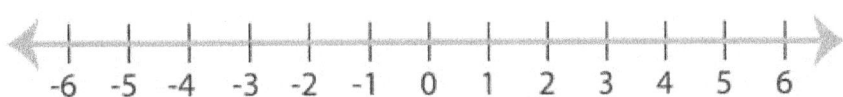

7) $9x \leq 27$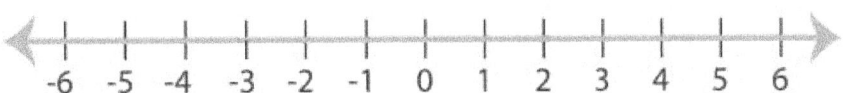

8) $x + 19 > 16$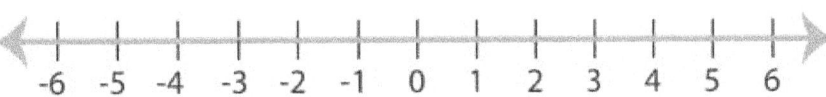

WWW.MathNotion.Com

ATI TEAS 6 Subject Test – Mathematics

# Multi-Step Inequalities

✎ **Calculate each inequality.**

1) $x - 3 \leq 7$

2) $8 - x \leq 8$

3) $3x - 9 \leq 9$

4) $4x - 4 \geq 8$

5) $x - 7 \geq 1$

6) $5x - 15 \leq 5$

7) $6x - 8 \leq 4$

8) $-11 + 6x \leq 12$

9) $4(x - 4) \leq 16$

10) $3x - 10 \leq 11$

11) $5x - 25 < 25$

12) $9x - 5 < 22$

13) $20 - 7x \geq -15$

14) $33 + 6x < 45$

15) $8 + 8x \geq 96$

16) $7 + 3x < 13$

17) $4x - 3 < 9$

18) $5(2 - 2x) \geq -30$

19) $-(7 + 6x) < 29$

20) $12 - 8x \geq -20$

21) $-4(x - 6) > 24$

22) $\dfrac{3x + 9}{6} \leq 10$

23) $\dfrac{4x - 10}{3} \leq 2$

24) $\dfrac{2x - 8}{3} > 2$

25) $8 + \dfrac{x}{6} < 9$

26) $\dfrac{9x}{7} - 4 < 5$

27) $\dfrac{15x + 45}{15} > 1$

28) $16 + \dfrac{x}{4} < 6$

WWW.MathNotion.Com

ATI TEAS 6 Subject Test – Mathematics

## Systems of Equations

✎ **Calculate each system of equations.**

1) $-x + y = 2$                    $x = \_\_\_$
   $-4x + 2y = 6$                  $y = \_\_\_$

2) $-15x + 3y = -9$                $x = \_\_\_$
   $9x - 16y = 48$                 $y = \_\_\_$

3) $y = -7$                        $x = \_\_\_$
   $6x + 5y = 7$                   $y = \_\_\_$

4) $3y = -9x + 15$                 $x = \_\_\_$
   $5x - 4y = -3$                  $y = \_\_\_$

5) $10x - 9y = -13$                $x = \_\_\_$
   $-5x + 3y = 11$                 $y = \_\_\_$

6) $-12x - 16y = 20$               $x = \_\_\_$
   $6x - 12y = 30$                 $y = \_\_\_$

7) $5x - 14y = -23$                $x = \_\_\_$
   $-18x + 21y = 24$               $y = \_\_\_$

8) $15x - 21y = -6$                $x = \_\_\_$
   $2x - 3y = -2$                  $y = \_\_\_$

9) $-x + 3y = 3$                   $x = \_\_\_$
   $-14x + 16y = -10$              $y = \_\_\_$

10) $x + 5y = 50$                  $x = \_\_\_$
    $3x + 10y = 80$                $y = \_\_\_$

11) $6x - 7y = -8$                 $x = \_\_\_$
    $-x - 4y = -9$                 $y = \_\_\_$

12) $2x + 4y = -10$                $x = \_\_\_$
    $2x - 8y = 14$                 $y = \_\_\_$

13) $4x + 3y = 12$                 $x = \_\_\_$
    $5x - 3y = 15$                 $y = \_\_\_$

14) $3x - 2y = 3$                  $x = \_\_\_$
    $7x - 8y = 22$                 $y = \_\_\_$

15) $3x + 2y = 5$                  $x = \_\_\_$
    $-10x - 4y = -14$              $y = \_\_\_$

16) $10x + 7y = 1$                 $x = \_\_\_$
    $-5x - 7y = 24$                $y = \_\_\_$

WWW.MathNotion.Com

ATI TEAS 6 Subject Test – Mathematics

# Systems of Equations Word Problems

✎ **Find the answer for each word problem.**

1) Tickets to a movie cost $4 for adults and $3 for students. A group of friends purchased 8 tickets for $31.00. How many adults ticket did they buy? ____

2) At a store, Eva bought two shirts and five hats for $77.00. Nicole bought three same shirts and four same hats for $84.00. What is the price of each shirt? _____

3) A farmhouse shelters 18 animals, some are pigs, and some are ducks. Altogether there are 66 legs. How many pigs are there? _____

4) A class of 214 students went on a field trip. They took 36 vehicles, some cars and some buses. If each car holds 5 students and each bus hold 22 students, how many buses did they take? _____

5) A theater is selling tickets for a performance. Mr. Smith purchased 5 senior tickets and 3 child tickets for $105 for his friends and family. Mr. Jackson purchased 3 senior tickets and 5 child tickets for $79. What is the price of a senior ticket? $_____

6) The difference of two numbers is 10. Their sum is 20. What is the bigger number? $_____

7) The sum of the digits of a certain two–digit number is 7. Reversing its digits increase the number by 9. What is the number? _____

8) The difference of two numbers is 11. Their sum is 25. What are the numbers? _____

9) The length of a rectangle is 5 meters greater than 2 times the width. The perimeter of rectangle is 28 meters. What is the length of the rectangle? _____

10) Jim has 25 nickels and dimes totaling $1.80. How many nickels does he have? _____

# ATI TEAS 6 Subject Test – Mathematics

## Answers of Worksheets

**One–Step Equations**

1) 30
2) 7
3) 4
4) 6
5) 5
6) 11
7) 11
8) 8
9) 17
10) −4
11) 12
12) 16
13) 34
14) −15
15) −18
16) 14
17) −14
18) 20
19) 45
20) −13
21) 34
22) −5
23) 47
24) −42
25) 22
26) 9
27) 60
28) 35
29) 54
30) 24
31) 30
32) −21

**Multi–Step Equations**

1) 2
2) −7
3) 4
4) 8
5) 6
6) 2
7) 10
8) 2
9) 2
10) −4
11) −4
12) −5
13) −7
14) 3
15) −2
16) 15
17) −2
18) −7
19) −5
20) −11
21) −4
22) −2
23) −7
24) 6
25) 3
26) 14
27) 7
28) 27
29) −5
30) 17
31) −6
32) −4

**Graphing Single–Variable Inequalities**

1)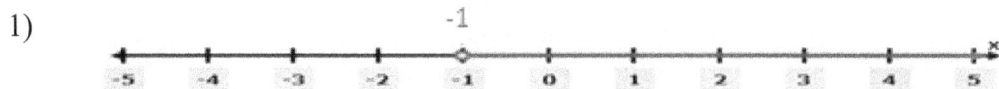

2)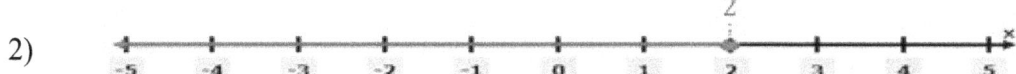

3)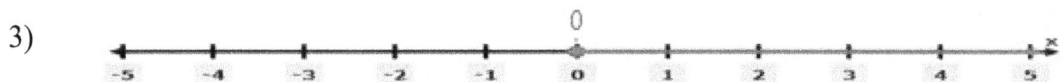

4)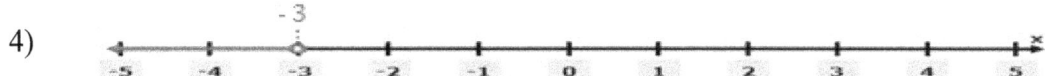

# ATI TEAS 6 Subject Test – Mathematics

5) [number line with open circle at 1/2, shaded right]

6) [number line with closed circle at -2, shaded right]

7) [number line with closed circle at 3, shaded right]

8) [number line with closed circle at -3.5, shaded left]

**One–Step Inequalities**

1) [number line with closed circle at 0, shaded left]

2) [number line with closed circle at 7, shaded left]

3) [number line with open circle at 7, shaded right]

4) [number line with closed circle at 2, shaded left]

5) [number line with open circle at -4, shaded left]

6) [number line with closed circle at 8, shaded right]

7) [number line with open circle at 3, shaded right]

8) [number line with open circle at -3, shaded right]

**Multi-Step Inequalities**

1) $x \leq 10$
2) $x \geq 0$
3) $x \leq 6$
4) $x \geq 3$
5) $x \geq 8$
6) $x \leq 4$
7) $x \leq 2$
8) $x \leq \frac{23}{6}$
9) $x \leq 8$
10) $x \leq 7$
11) $x < 10$
12) $x < 3$
13) $x \leq 5$
14) $x < 2$
15) $x \geq 11$
16) $x < 2$
17) $x < 3$
18) $x \leq 4$
19) $x > -6$
20) $x \leq 4$
21) $x < 0$
22) $x \leq 17$
23) $x \leq 4$

WWW.MathNotion.Com

# ATI TEAS 6 Subject Test – Mathematics

24) $x > 7$
25) $x < 6$
26) $x < 7$
27) $x > -2$
28) $x < -40$

**Systems of Equations**

1) $x = -1, y = 1$
2) $x = 0, y = -3$
3) $x = 7$
4) $x = 1, y = 2$
5) $x = -4, y = -3$
6) $x = 1, y = -2$
7) $x = 1, y = 2$
8) $x = 8, y = 6$
9) $x = 3, y = 2$
10) $x = -20, y = 14$
11) $x = 1, y = 2$
12) $x = -1, y = -2$
13) $x = 3, y = 0$
14) $x = -2, y = -\frac{9}{2}$
15) $x = 1, y = 1$
16) $x = 5, y = -7$

**Systems of Equations Word Problems**

1) 7
2) $16
3) 15
4) 2
5) $18
6) 15
7) 34
8) 18, 7
9) 11 meters
10) 14

# ATI TEAS 6 Subject Test – Mathematics

ATI TEAS 6 Subject Test – Mathematics

# Chapter 7:
# Linear Functions

**Topics that you'll practice in this chapter:**

- ✓ Finding Slope
- ✓ Graphing Lines Using Line Equation
- ✓ Writing Linear Equations
- ✓ Graphing Linear Inequalities
- ✓ Finding Midpoint
- ✓ Finding Distance of Two Points

*"Nature is written in mathematical language."* – Galileo Galilei

# ATI TEAS 6 Subject Test – Mathematics

## Finding Slope

✏️ **Find the slope of each line.**

1) $y = x + 8$

2) $y = -3x + 5$

3) $y = 2x + 12$

4) $y = -4x + 19$

5) $y = 11 + 6x$

6) $y = 7 - 5x$

7) $y = 8x + 19$

8) $y = -9x + 20$

9) $y = -7x + 4$

10) $y = 3x - 8$

11) $y = \frac{1}{3}x + 8$

12) $y = -\frac{4}{5}x + 9$

13) $-3x + 6y = 30$

14) $4x + 4y = 16$

15) $3y - x = 10$

16) $8y - x = 5$

✏️ **Find the slope of the line through each pair of points.**

17) $(2, 3), (7, 10)$

18) $(-3, 5), (2, 15)$

19) $(5, -3), (1, 9)$

20) $(-5, -5), (10, 25)$

21) $(22, 3), (7, 18)$

22) $(-16, 8), (-7, 26)$

23) $(25, 11), (29, 19)$

24) $(26, -19), (14, 17)$

25) $(22, -13), (20, -11)$

26) $(19, 7), (15, -3)$

27) $(5, 7), (11, 19)$

28) $(52, -62), (40, 70)$

WWW.MathNotion.Com

# ATI TEAS 6 Subject Test – Mathematics

## Graphing Lines Using Line Equation

✏️ Sketch the graph of each line.

1) $y = x - 2$

2) $y = -3x + 2$

3) $x + y = 0$

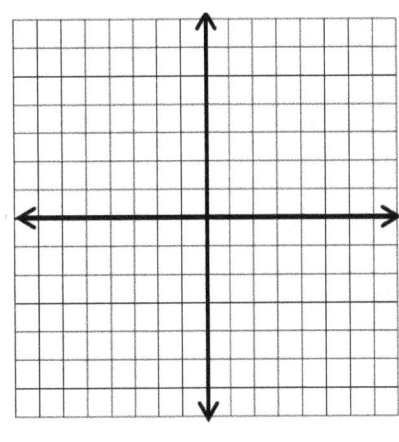
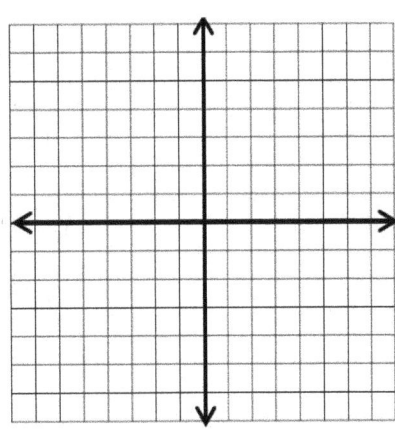
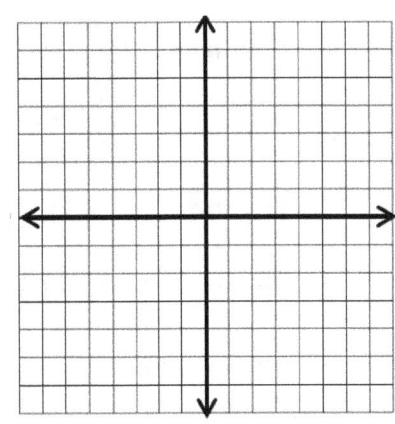

4) $x + y = -3$

5) $2x + 3y = -4$

6) $y - 3x + 6 = 0$

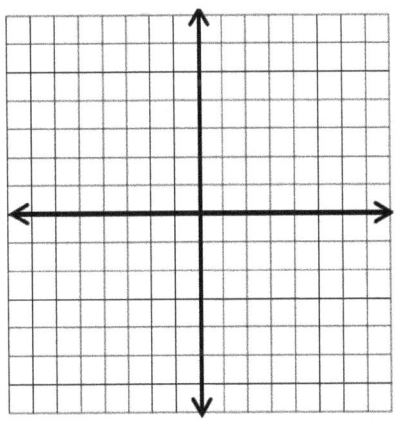
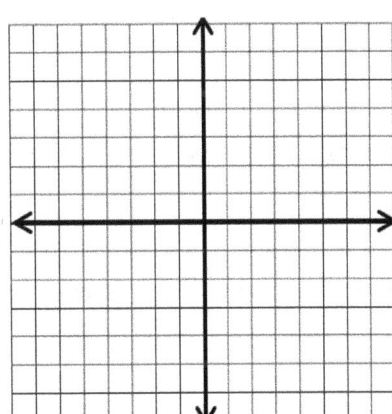
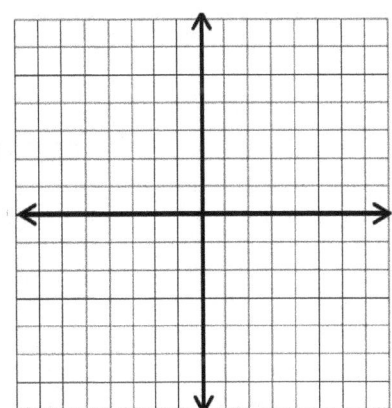

# ATI TEAS 6 Subject Test – Mathematics

## Writing Linear Equations

✍ **Write the equation of the line through the given points.**

1) Through: $(2, -5), (3, 9)$

2) Through: $(-6, 3), (3, 12)$

3) Through: $(10, 7), (5, 27)$

4) Through: $(15, 11), (3, -1)$

5) Through: $(24, 17), (12, -7)$

6) Through: $(8, 29), (4, -7)$

7) Through: $(20, -16), (12, 0)$

8) Through: $(-3, 10), (2, -5)$

9) Through: $(-6, 17), (4, -3)$

10) Through: $(-8, 22), (5, -4)$

11) Through: $(9, 27), (3, -3)$

12) Through: $(11, 32), (9, 4)$

13) Through: $(-3, 13), (-4, 0)$

14) Through: $(-5, 5), (5, 15)$

15) Through: $(18, -32), (11, 3)$

16) Through: $(-4, 25), (4, -15)$

✍ **Find the answer for each problem.**

17) What is the equation of a line with slope 6 and intercept 12? _____

18) What is the equation of a line with slope $-11$ and intercept $-4$? _____

19) What is the equation of a line with slope $-3$ and passes through point $(5, 2)$? _____

20) What is the equation of a line with slope $-5$ and passes through point $(-2, -1)$? _____

21) The slope of a line is $-10$ and it passes through point $(-3, 0)$. What is the equation of the line? _____

22) The slope of a line is 8 and it passes through point $(0, 7)$. What is the equation of the line? _____

# Graphing Linear Inequalities

✎ **Sketch the graph of each linear inequality.**

1) $y > 4x - 5$

2) $y < 2x + 4$

3) $y \leq -5x - 2$

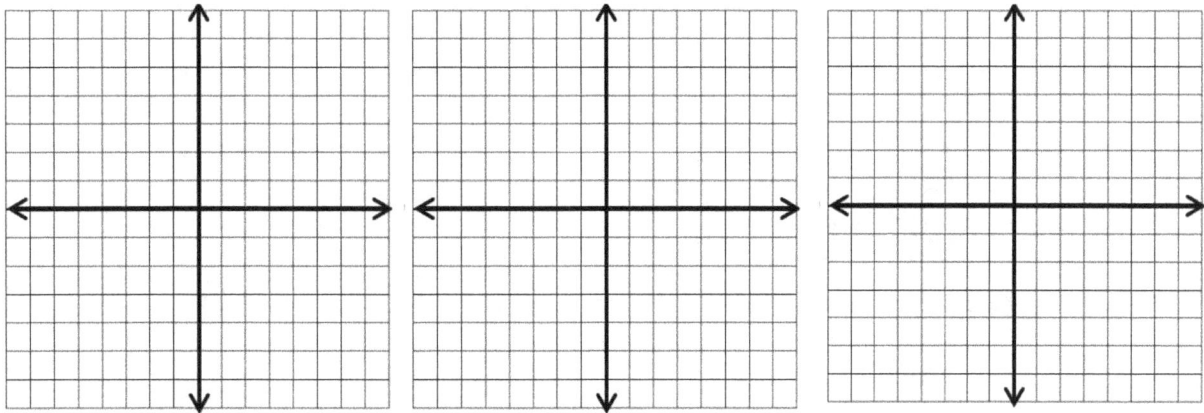

4) $4y \geq 12 + 4x$

5) $-12y < 3x - 24$

6) $5y \geq -15x + 10$

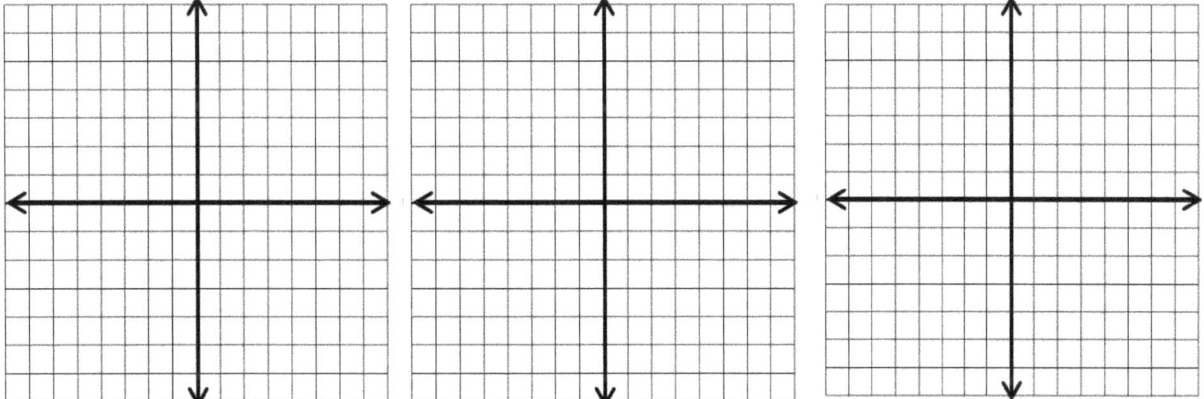

**ATI TEAS 6 Subject Test – Mathematics**

# Finding Midpoint

✎ Find the midpoint of the line segment with the given endpoints.

1) $(-4, -3), (2, 3)$

2) $(9, 0), (-1, 8)$

3) $(9, -6), (3, 14)$

4) $(-10, -6), (0, 8)$

5) $(2, -5), (14, -15)$

6) $(-10, -3), (4, -13)$

7) $(8, 7), (-8, 13)$

8) $(-3, 6), (-9, 2)$

9) $(-4, 5), (16, -9)$

10) $(7, 14), (9, -2)$

11) $(-8, 6), (6, 6)$

12) $(10, 5), (-2, -3)$

13) $(-5, 12), (-3, 3)$

14) $(12, 7), (8, -2)$

15) $(10, 2), (-6, 14)$

16) $(-1, -2), (-7, 10)$

17) $(7, -7), (13, -13)$

18) $(-3, -8), (11, -4)$

19) $(5, -11), (-8, 9)$

20) $(14, -4), (16, 14)$

21) $(0, -5), (8, -1)$

22) $(3, 0), (-21, 18)$

23) $(17, -3), (-7, -5)$

24) $(26, -12), (6, 24)$

✎ Find the answer for each problem.

25) One endpoint of a line segment is $(-3, 7)$ and the midpoint of the line segment is $(-6, 9)$. What is the other endpoint? _____

26) One endpoint of a line segment is $(-3, 7)$ and the midpoint of the line segment is $(1, 5)$. What is the other endpoint? _____

27) One endpoint of a line segment is $(-10, -16)$ and the midpoint of the line segment is $(2, 9)$. What is the other endpoint? _____

# ATI TEAS 6 Subject Test – Mathematics

## Finding Distance of Two Points

✍ Find the distance between each pair of points.

1) $(6, 3), (-3, -9)$

2) $(5, 2), (-10, -6)$

3) $(8, 5), (8, 3)$

4) $(-8, -2), (2, 22)$

5) $(6, -7), (-3, -7)$

6) $(12, 0), (-9, -20)$

7) $(3, 20), (3, -5)$

8) $(10, 17), (5, 5)$

9) $(7, -2), (-4, -2)$

10) $(13, 4), (5, -2)$

11) $(11, 13), (5, 5)$

12) $(1, 4), (-23, -3)$

13) $(9, 8), (5, -4)$

14) $(-11, -4), (5, 8)$

15) $(-2, -6), (-2, -12)$

16) $(-1, -4), (23, 3)$

17) $(19, 3), (7, -6)$

18) $(-5, -2), (3, 4)$

19) $(2, 6), (2, -12)$

20) $(-4, -2), (8, -2)$

✍ Find the answer for each problem.

21) Triangle ABC is a right triangle on the coordinate system and its vertices are $(-2, 5)$, $(-2, 1)$, and $(1, 1)$. What is the area of triangle ABC? _____

22) Three vertices of a triangle on a coordinate system are $(3, -6)$, $(-5, -12)$, and $(3, -18)$. What is the perimeter of the triangle? _____

23) Four vertices of a rectangle on a coordinate system are $(-2, 2)$, $(-2, 6)$, $(4, 2)$, and $(4, 6)$. What is its perimeter? _____

WWW.MathNotion.Com

# ATI TEAS 6 Subject Test – Mathematics

## Answers of Worksheets

**Finding Slope**

1) 1
2) $-3$
3) 2
4) $-4$
5) 6
6) $-5$
7) 8
8) $-9$
9) $-7$
10) 3
11) $\frac{1}{3}$
12) $-\frac{4}{5}$
13) $\frac{1}{2}$
14) $-1$
15) $\frac{1}{3}$
16) $\frac{1}{8}$
17) $\frac{7}{5}$
18) 2
19) $-3$
20) 2
21) $-1$
22) 2
23) 2
24) $-3$
25) $-1$
26) $\frac{5}{2}$
27) 2
28) $-11$

**Graphing Lines Using Line Equation**

1) $y = x - 2$

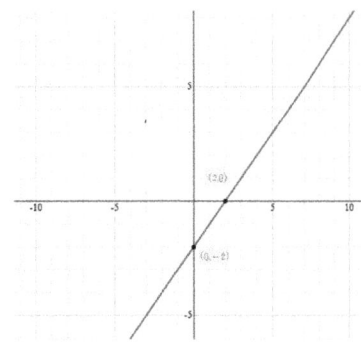

2) $y = -3x + 2$

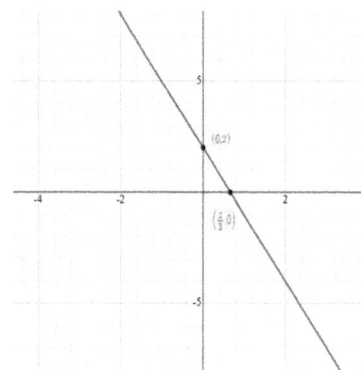

3) $x + y = 0$

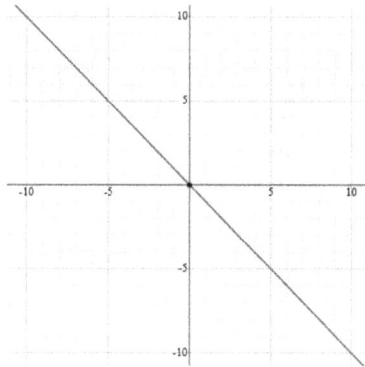

4) $x + y = -3$

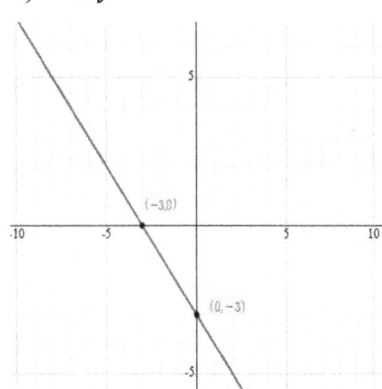

5) $2x + 3y = -4$

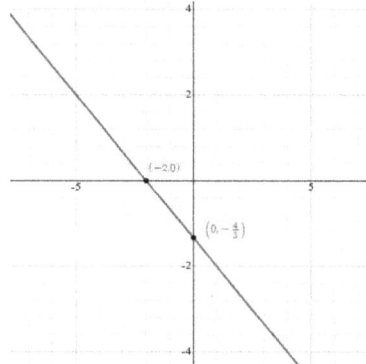

6) $y - 3x + 6 = 0$

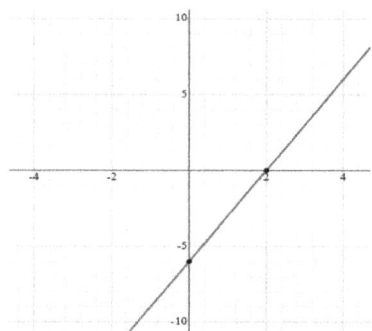

# ATI TEAS 6 Subject Test – Mathematics

**Writing Linear Equations**

1) $y = 14x - 33$
2) $y = x + 9$
3) $y = -4x + 47$
4) $y = x - 4$
5) $y = 2x - 31$
6) $y = 9x - 43$
7) $y = -2x + 24$
8) $y = -3x + 1$
9) $y = -2x + 5$
10) $y = -2x + 6$
11) $y = 5x - 18$
12) $y = 14x - 122$
13) $y = 13x + 52$
14) $y = x + 10$
15) $y = -5x + 58$
16) $y = -5x + 5$
17) $y = 6x + 12$
18) $y = -11x - 4$
19) $y = -3x + 17$
20) $y = -5x - 11$
21) $y = -10x - 30$
22) $y = 8x + 7$

**Graphing Linear Inequalities**

1) $y > 4x - 5$

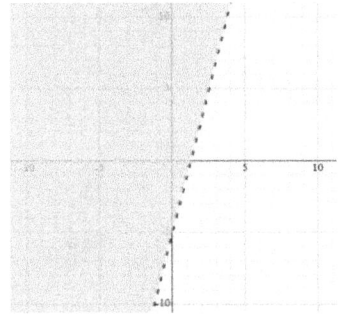

2) $y < 2x + 4$

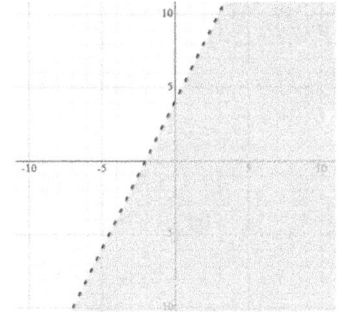

3) $y \leq -5x - 2$

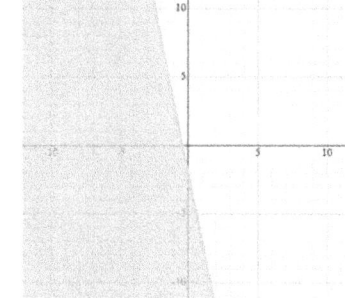

4) $4y \geq 12 + 4x$

5) $-12y < 3x - 24$

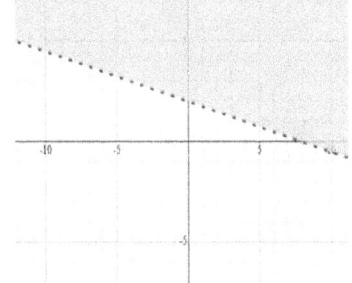

6) $5y \geq -15x + 10$

**Finding Midpoint**

1) $(-1, 0)$
2) $(4, 4)$
3) $(6, 4)$
4) $(-5, 1)$
5) $(8, -10)$
6) $(-3, -8)$
7) $(0, 10)$
8) $(-6, 4)$
9) $(6, -2)$
10) $(8, 6)$
11) $(-1, 6)$
12) $(4, 1)$
13) $(-4, 7.5)$
14) $(10, 2.5)$
15) $(2, 8)$
16) $(-4, 4)$
17) $(10, -10)$
18) $(4, -6)$

# ATI TEAS 6 Subject Test – Mathematics

19) $(-1.5, -1)$
20) $(15, 5)$
21) $(4, -3)$

22) $(-9, 9)$
23) $(5, -4)$
24) $(16, 6)$

25) $(-9, 11)$
26) $(5, 3)$
27) $(14, 34)$

**Finding Distance of Two Points**

1) 15
2) 17
3) 2
4) 26
5) 9
6) 29
7) 25
8) 13

9) 11
10) 10
11) 10
12) 25
13) $4\sqrt{10}$
14) 20
15) 6
16) 25

17) 15
18) 10
19) 18
20) 12
21) 6 *square units*
22) 32 *units*
23) 20 *units*

ATI TEAS 6 Subject Test – Mathematics

# Chapter 8:

# Polynomials

## Topics that you'll practice in this chapter:

- ✓ Writing Polynomials in Standard Form
- ✓ Simplifying Polynomials
- ✓ Adding and Subtracting Polynomials
- ✓ Multiplying Monomials
- ✓ Multiplying and Dividing Monomials
- ✓ Multiplying a Polynomial and a Monomial
- ✓ Multiplying Binomials
- ✓ Factoring Trinomials
- ✓ Operations with Polynomials

*Mathematics is the supreme judge; from its decisions there is no appeal.* – Tobias Dantzig

ATI TEAS 6 Subject Test – Mathematics

## Writing Polynomials in Standard Form

✎ Write each polynomial in standard form.

1) $11x - 7x =$

2) $-5 + 19x - 19x =$

3) $6x^5 - 12x^3 =$

4) $12 + 17x^4 - 12 =$

5) $5x^2 + 4x - 9x^3 =$

6) $-3x^2 + 12x^5 =$

7) $5x + 8x^3 - 2x^8 =$

8) $-7x^3 + 4x - 9x^6 =$

9) $3x^2 + 22 - 6x =$

10) $3 - 4x + 9x^4 =$

11) $13x^2 + 28x - 8x^3 =$

12) $16 + 4x^2 - 2x^3 =$

13) $19x^2 - 9x + 9x^4 =$

14) $3x^4 - 7x^2 - 2x^3 =$

15) $-51 + 3x^2 - 8x^4 =$

16) $7x^2 - 8x^6 + 4x^4 - 15 =$

17) $6x^4 - 4x^5 + 16 - 3x^3 =$

18) $-2x^6 + 4x - 7x^2 - 5x =$

19) $11x^7 + 8x^5 - 5x^7 - 3x^2 =$

20) $2x^2 - 12x^5 + 8x^2 + 3x^6 =$

21) $4x^5 - 11x^7 - 6x^3 + 16x^5 =$

22) $6x^3 + 3x^5 + 34x^4 - 8x^5 =$

23) $3x(4x + 5 - 2x^2) =$

24) $12x(x^6 + 4x^3) =$

25) $5x(3x^2 + 6x + 4) =$

26) $7x(4 - 2x + 6x^5) =$

27) $3x(4x^4 - 4x^3 + 2) =$

28) $4x(2x^5 + 6x^2 - 3) =$

29) $5x(3x^4 + 4x^3 + 2x) =$

30) $2x(3x - 2x^3 + 4x^6) =$

WWW.MathNotion.Com

ATI TEAS 6 Subject Test – Mathematics

# Simplifying Polynomials

✎ **Simplify each expression.**

1) $3(4x - 20) =$

2) $5x(3x - 4) =$

3) $6x(5x - 7) =$

4) $3x(7x + 5) =$

5) $5x(4x - 3) =$

6) $6x(8x + 2) =$

7) $(3x - 2)(x - 4) =$

8) $(x - 5)(2x + 6) =$

9) $(x - 3)(x - 7) =$

10) $(3x + 4)(3x - 4) =$

11) $(5x - 4)(5x - 2) =$

12) $6x^2 + 6x^2 - 8x^4 =$

13) $3x - 2x^2 + 5x^3 + 7 =$

14) $7x + 4x^2 - 10x^3 =$

15) $12x^2 + 5x^5 - 6x^3 =$

16) $-5x^2 + 4x^6 + 6x^8 =$

17) $-12x^3 + 10x^5 - 4x^6 + 4x =$

18) $11 - 7x^2 + 4x^2 - 16x^3 + 11 =$

19) $2x^2 - 9x + 4x^3 + 15x - 10x =$

20) $13 - 7x^5 + 6x^5 - 4x^2 + 5 =$

21) $-5x^8 + x^6 - 14x^3 + 5x^8 =$

22) $(7x^4 - 4) + (7x^4 - 2x^4) =$

23) $3(3x^4 - 4x^3 - 6x^4) =$

24) $-5(x^9 + 8) - 5(10 - x^9) =$

25) $8x^3 - 9x^4 - 2x + 19 - 8x^3 =$

26) $11 - 8x^3 + 6x^3 - 7x^5 + 6 =$

27) $(5x^3 - 4x) - (6x - 2 - 6x^3) =$

28) $4x^2 - 5x^4 - x(3x^3 + 2x) =$

29) $6x + 6x^5 - 10 - 4(x^5 - 3) =$

30) $4 - 3x^4 + (6x^5 - 2x^4 + 5x^5) =$

31) $-(x^5 + 4) - 8(3 + x^5) =$

32) $(4x^3 - 3x) - (3x - 5x^3) =$

ATI TEAS 6 Subject Test – Mathematics

## Adding and Subtracting Polynomials

✎ **Add or subtract expressions.**

1) $(-2x^2 - 3) + (3x^2 + 4) =$

2) $(4x^3 + 6) - (7 - 2x^3) =$

3) $(4x^5 + 5x^2) - (2x^5 + 15) =$

4) $(6x^3 - 2x^2) + (5x^2 - 4x) =$

5) $(10x^4 + 28x) - (34x^4 + 6) =$

6) $(7x^2 - 3) + (7x^2 + 3) =$

7) $(9x^2 + 4) - (10 - 5x^2) =$

8) $(6x^2 + x^5) - (x^5 + 4) =$

9) $(4x^3 - x) + (3x - 7x^3) =$

10) $(11x + 10) - (8x + 10) =$

11) $(15x^3 - 3x) - (3x - 4x^3) =$

12) $(4x - x^5) - (6x^5 + 8x) =$

13) $(2x^2 - 7x^7) - (4x^7 - 6x) =$

14) $(3x^2 - 5) + (8x^2 + 4x^5) =$

15) $(9x^4 + 5x^5) - (x^5 - 9x^4) =$

16) $(-4x^3 - 2x) + (9x - 5x^3) =$

17) $(4x - 3x^2) - (148x^2 + x) =$

18) $(5x - 8x^4) - (3x^4 - 4x^2) =$

19) $(8x^4 - 4) + (2x^4 - 3x^2) =$

20) $(5x^6 + 7x^3) - (x^3 - 5x^6) =$

21) $(-2x^2 + 20x^5 + 5x^4) + (12x^4 + 8x^5 + 24x^2) =$

22) $(7x^4 - 9x^7 - 6x) - (-3x^4 - 9x^7 + 6x) =$

23) $(14x + 12x^4 - 18x^6) + (20x^4 + 18x^6 - 10x) =$

24) $(5x^8 - 6x^6 - 4x) - (5x^3 + 9x^6 - 7x) =$

25) $(11x^2 - 6x^4 - 3x) - (-4x^2 - 12x^4 + 9x) =$

26) $(-5x^9 + 14x^3 + 3x^7) + (10x^7 + 26x^3 + 3x^9) =$

WWW.MathNotion.Com

ATI TEAS 6 Subject Test – Mathematics

# Multiplying Monomials

**Simplify each expression.**

1) $6u^8 \times (-u^2) =$

2) $(-5p^8) \times (-2p^3) =$

3) $4xy^3z^5 \times 3z^4 =$

4) $3u^5t \times 8ut^4 =$

5) $(-5a^2) \times (-7a^3b^6) =$

6) $-3a^4b^3 \times 6a^2b =$

7) $13xy^5 \times x^4y^4 =$

8) $6p^4q^3 \times (-8pq^6) =$

9) $8s^4t^3 \times 4st^3 =$

10) $(-6x^4y^3) \times 6x^2y =$

11) $3xy^7z \times 12z^3 =$

12) $24xy \times x^2y =$

13) $13pq^4 \times (-3p^2q) =$

14) $13s^3t^4 \times st^4 =$

15) $11p^5 \times (-6p^3) =$

16) $(-8p^3q^5r) \times 3pq^4r^6 =$

17) $(-4a^4) \times (-7a^3b) =$

18) $6u^6v^2 \times (-5u^3v^4) =$

19) $9u^5 \times (-3u) =$

20) $-6xy^5 \times 4x^2y =$

21) $13y^5z^3 \times (-y^3z) =$

22) $8a^4bc^3 \times 2abc^3 =$

23) $(-7p^5q^6) \times (-5p^4q^2) =$

24) $4u^5v^3 \times (-4u^7v^3) =$

25) $17y^4z^5 \times (-y^6z) =$

26) $(-5pq^3r^2) \times 8p^2q^4r =$

27) $3ab^5c^6 \times 5a^4bc^2 =$

28) $6x^3yz^2 \times 3x^2y^7z^3 =$

ATI TEAS 6 Subject Test – Mathematics

## Multiplying and Dividing Monomials

✏️ **Simplify each expression.**

1) $(5x^5)(2x^2) =$

2) $(4x^4)(6x^2) =$

3) $(3x^4)(7x^4) =$

4) $(5x^6)(4x^2) =$

5) $(12x^4)(3x^6) =$

6) $(4yx^8)(8y^4x^3) =$

7) $(14x^4y)(x^3y^5) =$

8) $(-5x^3y^4)(2x^3y^5) =$

9) $(-6x^4y^2)(-3x^3y^5) =$

10) $(5x^3y)(-5x^2y^3) =$

11) $(6x^4y^3)(4x^3y^4) =$

12) $(4x^3y^2)(5x^2y^4) =$

13) $(12x^3y^6)(4x^4y^{10}) =$

14) $(15x^3y^5)(3x^4y^6) =$

15) $(7x^2y^7)(8x^6y^7) =$

16) $(-3x^3y^8)(7x^9y^4) =$

17) $\dfrac{5x^6y^6}{xy^4} =$

18) $\dfrac{19x^7y^5}{19x^6y} =$

19) $\dfrac{56x^4y^4}{8xy} =$

20) $\dfrac{81x^5y^6}{9x^4y^5} =$

21) $\dfrac{36x^7y^6}{9x^2y^3} =$

22) $\dfrac{48x^9y^7}{4x^4y^6} =$

23) $\dfrac{88x^{18}y^{12}}{11x^8y^9} =$

24) $\dfrac{30x^7y^6}{6x^8y^3} =$

25) $\dfrac{150x^7y^6}{30x^4y^6} =$

26) $\dfrac{-42x^{18}y^{14}}{6x^4y^9} =$

27) $\dfrac{-36x^7y^8}{9x^5y^8} =$

WWW.MathNotion.Com

# Multiplying a Polynomial and a Monomial

✏️ **Find each product.**

1) $x(2x + 4) =$

2) $6(4 - 2x) =$

3) $5x(4x + 2) =$

4) $x(-4x + 5) =$

5) $8x(2x - 2) =$

6) $6(2x - 4y) =$

7) $7x(5x - 5) =$

8) $3x(12x + 2y) =$

9) $4x(x + 6y) =$

10) $11x(3x + 4y) =$

11) $7x(3x + 2) =$

12) $10x(4x - 10y) =$

13) $9x(3x - 2y) =$

14) $7x(x - 4y + 6) =$

15) $8x(2x^2 + 5y^2) =$

16) $12x(2x + 3y) =$

17) $4(2x^4 - 4y^4) =$

18) $4x(-3x^2y + 4y) =$

19) $-4(5x^3 - 2xy + 4) =$

20) $4(x^2 - 5xy - 6) =$

21) $8x(2x^3 - 5xy + 2x) =$

22) $-6x(-2x^3 - 6x + 2xy) =$

23) $3(2x^2 + xy - 9y^2) =$

24) $4x(5x^3 - 3x + 7) =$

25) $6(3x^{22} - 2x - 5) =$

26) $x^2(-2x^3 + 4x + 3) =$

27) $x^2(4x^3 + 10 - 2x) =$

28) $4x^4(3x^3 - 2x + 5) =$

29) $2x^2(4x^4 - 5xy + 7y^3) =$

30) $5x^2(5x^4 - 3x + 9) =$

31) $7x^2(6x^2 + 3x - 6) =$

32) $4x(x^3 - 4xy + 2y^2) =$

# ATI TEAS 6 Subject Test – Mathematics

## Multiplying Binomials

**✏ Find each product.**

1) $(x + 3)(x + 6) =$

2) $(x - 4)(x + 3) =$

3) $(x - 3)(x - 8) =$

4) $(x + 8)(x + 9) =$

5) $(x - 2)(x - 12) =$

6) $(x + 5)(x + 5) =$

7) $(x - 6)(x + 7) =$

8) $(x - 8)(x - 3) =$

9) $(x + 7)(x + 12) =$

10) $(x - 4)(x + 8) =$

11) $(x + 8)(x + 8) =$

12) $(x + 2)(x + 7) =$

13) $(x - 6)(x + 6) =$

14) $(x - 5)(x + 5) =$

15) $(x + 11)(x + 11) =$

16) $(x + 6)(x + 9) =$

17) $(x - 2)(x + 2) =$

18) $(x - 4)(x + 7) =$

19) $(3x + 5)(x + 6) =$

20) $(5x - 6)(4x + 8) =$

21) $(x - 7)(3x + 7) =$

22) $(x - 9)(x - 4) =$

23) $(x - 12)(x + 2) =$

24) $(2x - 4)(5x + 4) =$

25) $(3x - 8)(x + 8) =$

26) $(7x - 2)(6x + 3) =$

27) $(4x + 5)(3x + 5) =$

28) $(7x - 4)(9x + 4) =$

29) $(x + 2)(2x - 8) =$

30) $(5x - 4)(5x + 4) =$

31) $(3x + 2)(3x - 7) =$

32) $(x^2 + 8)(x^2 - 8) =$

# ATI TEAS 6 Subject Test – Mathematics

## Factoring Trinomials

✎ **Factor each trinomial.**

1) $x^2 + 8x + 12 =$

2) $x^2 - 6x + 5 =$

3) $x^2 + 15x + 36 =$

4) $x^2 - 12x + 35 =$

5) $x^2 - 11x + 18 =$

6) $x^2 - 9x + 18 =$

7) $x^2 + 18x + 72 =$

8) $x^2 - x - 72 =$

9) $x^2 + 4x - 21 =$

10) $x^2 - 13x + 22 =$

11) $x^2 + 2x - 24 =$

12) $x^2 - 3x - 40 =$

13) $x^2 - 3x - 70 =$

14) $x^2 + 26x + 169 =$

15) $4x^2 - 7x - 15 =$

16) $x^2 - 14x + 33 =$

17) $10x^2 + 5x - 15 =$

18) $6x^2 - 4x - 42 =$

19) $x^2 + 12x + 36 =$

20) $5x^2 + 17x - 12 =$

✎ **Calculate each problem.**

21) The area of a rectangle is $x^2 - x - 56$. If the width of rectangle is $x + 7$, what is its length? _____

22) The area of a parallelogram is $4x^2 + 17x - 15$ and its height is $x + 5$. What is the base of the parallelogram? _____

23) The area of a rectangle is $6x^2 - 22x + 12$. If the width of the rectangle is $3x - 2$, what is its length? _____

ATI TEAS 6 Subject Test – Mathematics

# Operations with Polynomials

✏ **Find each product.**

1) $4(5x + 3) =$ _____

2) $8(2x + 6) =$ _____

3) $2(5x - 2) =$ _____

4) $-4(7x - 3) =$ _____

5) $3x^2(9x + 1) =$ _____

6) $4x^6(7x - 9) =$ _____

7) $3x^4(-7x + 3) =$ _____

8) $-8x^4(5x - 8) =$ _____

9) $7(x^2 + 5x - 3) =$ _____

10) $9(5x^2 - 7x + 5) =$ _____

11) $3(3x^2 + 3x + 2) =$ _____

12) $5x(3x^2 + 5x + 8) =$ _____

13) $(5x + 7)(3x - 3) =$ _____

14) $(9x + 3)(3x - 5) =$ _____

15) $(6x + 3)(4x - 2) =$ _____

16) $(7x - 2)(3x + 5) =$ _____

✏ **Calculate each problem.**

17) The measures of two sides of a triangle are $(2x + 5y)$ and $(6x - 3y)$. If the perimeter of the triangle is $(13x + 4y)$, what is the measure of the third side? _____

18) The height of a triangle is $(8x + 5)$ and its base is $(4x - 3)$. What is the area of the triangle? _____

19) One side of a square is $(6x + 2)$. What is the area of the square? _____

20) The length of a rectangle is $(5x - 8y)$ and its width is $(15x + 8y)$. What is the perimeter of the rectangle? _____

21) The side of a cube measures $(x + 2)$. What is the volume of the cube? _____

22) If the perimeter of a rectangle is $(28x + 6y)$ and its width is $(5x + 2y)$, what is the length of the rectangle? _____

WWW.MathNotion.Com

# ATI TEAS 6 Subject Test – Mathematics

## Answers of Worksheets

**Writing Polynomials in Standard Form**

1) $4x$
2) $-5$
3) $6x^5 - 12x^3$
4) $14x^4$
5) $-9x^3 + 5x^2 + 4x$
6) $12x^5 - 3x^2$
7) $-2x^8 + 8x^3 + 5x$
8) $-9x^6 - 7x^3 + 4x$
9) $3x^2 - 6x + 22$
10) $9x^4 - 4x + 3$
11) $-8x^3 + 13x^2 + 28x$
12) $-2x^3 + 4x^2 + 16$
13) $9x^4 + 19x^2 - 9x$
14) $3x^4 - 2x^3 - 7x^2$
15) $-8x^4 + 3x^2 - 51$
16) $-8x^6 + 4x^4 + 7x^2 - 15$
17) $-4x^5 + 6x^4 - 3x^3 + 16$
18) $-2x^6 - 7x^2 - x$
19) $6x^7 + 8x^5 - 3x^2$
20) $3x^6 - 12x^5 + 10x^2$
21) $-11x^7 + 20x^5 - 6x^3$
22) $-5x^5 + 34x^4 + 6x^3$
23) $-6x^3 + 12x^2 + 15x$
24) $12x^7 + 48x^4$
25) $15x^3 + 30x^2 + 20x$
26) $42x^6 - 14x^2 + 28x$
27) $12x^5 - 12x^4 + 6x$
28) $8x^6 + 24x^3 - 12x$
29) $15x^5 + 20x^4 + 10x^2$
30) $8x^7 - 4x^4 + 6x^2$

**Simplifying Polynomials**

1) $12x - 60$
2) $15x^2 - 20x$
3) $30x^2 - 42x$
4) $21x^2 + 15x$
5) $20x^2 - 15x$
6) $48x^2 + 12x$
7) $3x^2 - 14x + 8$
8) $2x^2 - 4x - 30$
9) $x^2 - 10x + 21$
10) $9x^2 - 16$
11) $25x^2 - 30x + 8$
12) $-8x^4 + 12x^2$
13) $5x^3 - 2x^2 + 3x + 7$
14) $-10x^3 + 4x^2 + 7x$
15) $5x^5 - 6x^3 + 12x^2$
16) $6x^8 + 4x^6 - 5x^2$
17) $-4x^6 + 10x^5 - 12x^3 + 4x$
18) $-16x^3 - 3x^2 + 22$
19) $4x^3 + 2x^2 - 4x$
20) $-x^5 - 4x^2 + 18$
21) $x^6 - 14x^3$
22) $12x^4 - 4$
23) $-9x^4 - 12x^3$
24) $-90$

WWW.MathNotion.Com

# ATI TEAS 6 Subject Test – Mathematics

25) $-9x^4 - 2x + 19$

26) $-7x^5 - 2x^3 + 17$

27) $11x^3 - 10x + 2$

28) $-8x^4 + 2x^2$

29) $2x^5 + 6x + 2$

30) $11x^5 - 5x^4 + 4$

31) $-9x^5 - 28$

32) $9x^3 - 6x$

## Adding and Subtracting Polynomials

1) $x^2 + 1$

2) $6x^3 - 1$

3) $2x^5 + 5x^2 - 15$

4) $6x^3 + 3x^2 - 4x$

5) $-24x^4 + 28x - 6$

6) $14x^2$

7) $14x^2 - 6$

8) $6x^2 - 4$

9) $-3x^3 + 2x$

10) $3x$

11) $19x^3 - 6x$

12) $-7x^5 - 4x$

13) $-11x^7 + 2x^2 + 6x$

14) $4x^5 + 11x^2 - 5$

15) $4x^5 + 18x^4$

16) $-9x^3 + 7x$

17) $-151x^2 + 3x$

18) $-11x^4 + 4x^2 + 5x$

19) $10x^4 - 3x^2 - 4$

20) $10x^6 + 6x^3$

21) $28x^5 + 17x^4 + 22x^2$

22) $10x^4 - 12x$

23) $32x^4 + 4x$

24) $5x^8 - 15x^6 - 5x^3 + 3x$

25) $6x^4 + 15x^2 - 12x$

26) $-2x^9 + 13x^7 + 40x^3$

## Multiplying Monomials

1) $-6u^{10}$

2) $10p^{11}$

3) $12xy^3z^9$

4) $24u^6t^5$

5) $35a^5b^6$

6) $-18a^6b^4$

7) $13x^5y^9$

8) $-48p^5q^9$

9) $32s^5t^6$

10) $-36x^6y^4$

11) $36xy^7z^4$

12) $24px^3y^2$

13) $-39p^3q^5$

14) $13s^4t^8$

15) $-66p^8$

16) $-24p^4q^9r^7$

17) $28a^7b$

18) $-30u^9v^6$

19) $-27u^6$

20) $-24x^3y^6$

21) $-13y^8z^4$

22) $16a^5b^2c^6$

23) $35p^9q^8$

24) $-16u^{12}v^6$

25) $-17y^{10}z^6$

26) $-40p^3q^7r^3$

27) $15a^5b^6c^8$

28) $18x^5y^8z^5$

## Multiplying and Dividing Monomials

1) $10x^7$

2) $24x^6$

3) $21x^8$

4) $20x^8$

5) $36x^{10}$

6) $32x^{11}y^5$

7) $14x^7y^6$

8) $-10x^6y^9$

9) $18x^7y^7$

10) $-25x^5y^4$

11) $24x^7y^7$

12) $20x^5y^6$

# ATI TEAS 6 Subject Test – Mathematics

13) $48x^7y^{16}$
14) $45x^7y^{11}$
15) $56x^8y^{14}$
16) $-21x^{12}y^{12}$
17) $5x^5y^2$
18) $xy^4$
19) $7x^3y^3$
20) $9xy$
21) $4x^5y^3$
22) $12x^5y$
23) $8x^{10}y^3$
24) $5x^{-1}y^3$
25) $5x^3$
26) $-7x^{14}y^5$
27) $-4x^2$

## Multiplying a Polynomial and a Monomial

1) $2x^2 + 4x$
2) $-12x + 24$
3) $20x^2 + 10x$
4) $-4x^2 + 5x$
5) $16x^2 - 16x$
6) $12x - 24y$
7) $35x^2 - 35x$
8) $36x^2 + 6xy$
9) $4x^2 + 24xy$
10) $33x^2 + 44xy$
11) $21x^2 + 14x$
12) $40x^2 - 100xy$
13) $27x^2 - 18xy$
14) $7x^2 - 28xy + 42x$
15) $16x^3 + 40xy^2$
16) $24x^2 + 36xy$
17) $8x^4 - 16y^4$
18) $-12x^3y + 16xy$
19) $-20x^3 + 8xy - 16$
20) $4x^2 - 20xy - 24$
21) $16x^4 - 40x^2y + 16x^2$
22) $12x^4 + 36x^2 - 12x^2y$
23) $6x^2 + 3xy - 27y^2$
24) $20x^4 - 12x^2 + 28x$
25) $18x^{22} - 12x - 30$
26) $-2x^5 + 4x^3 + 3x^2$
27) $4x^5 - 2x^3 + 10x^2$
28) $12x^7 - 8x^5 + 20x^4$
29) $8x^6 - 10x^3y + 14x^2y^3$
30) $25x^6 - 15x^3 + 45x^2$
31) $42x^4 + 21x^3 - 42x^2$
32) $4x^4 - 16x^2y + 8xy^2$

## Multiplying Binomials

1) $x^2 + 9x + 18$
2) $x^2 - x - 12$
3) $x^2 - 11x + 24$
4) $x^2 + 17x + 72$
5) $x^2 - 14x + 24$
6) $x^2 + 10x + 25$
7) $x^2 + x - 42$
8) $x^2 - 11x + 24$
9) $x^2 + 19x + 84$
10) $x^2 + 4x - 32$
11) $x^2 + 16x + 64$
12) $x^2 + 9x + 14$
13) $x^2 - 36$
14) $x^2 - 25$

# ATI TEAS 6 Subject Test – Mathematics

15) $x^2 + 22x + 121$
16) $x^2 + 15x + 54$
17) $x^2 - 4$
18) $x^2 + 3x - 28$
19) $3x^2 + 23x + 30$
20) $20x^2 + 16x - 48$
21) $3x^2 - 14x - 49$
22) $x^2 - 13x + 36$
23) $x^2 - 10x - 24$

24) $10x^2 - 12x - 16$
25) $3x^2 + 16x - 64$
26) $42x^2 + 9x - 6$
27) $12x^2 + 35x + 25$
28) $63x^2 - 8x - 16$
29) $2x^2 - 4x - 16$
30) $25x^2 - 16$
31) $9x^2 - 15x - 14$
32) $x^4 - 64$

**Factoring Trinomials**

1) $(x + 6)(x + 2)$
2) $(x - 5)(x - 1)$
3) $(x + 12)(x + 3)$
4) $(x - 5)(x - 7)$
5) $(x - 2)(x - 9)$
6) $(x - 6)(x - 3)$
7) $(x + 6)(x + 12)$
8) $(x + 8)(x - 9)$

9) $(x - 3)(x + 7)$
10) $(x - 11)(x - 2)$
11) $(x - 4)(x + 6)$
12) $(x - 8)(x + 5)$
13) $(x + 7)(x - 10)$
14) $(x + 13)(x + 13)$
15) $(4x + 5)(x - 3)$
16) $(x - 11)(x - 3)$

17) $(5x - 5)(2x + 3)$
18) $(2x - 6)(3x + 7)$
19) $(x + 6)(x + 6)$
20) $(5x - 3)(x + 4)$
21) $(x - 8)$
22) $(4x - 3)$
23) $(2x - 6)$

**Operations with Polynomials**

1) $20x + 12$
2) $16x + 48$
3) $10x - 4$
4) $-28x + 12$
5) $27x^3 + 3x^2$
6) $28x^7 - 36x^6$
7) $-21x^5 + 9x^4$
8) $-40x^5 + 64x^4$

9) $7x^2 + 35x - 21$
10) $45x^2 - 63x + 45$
11) $9x^2 + 9x + 6$
12) $15x^3 + 25x^2 + 40x$
13) $15x^2 + 6x - 21$
14) $27x^2 - 36x - 15$
15) $24x^2 - 6$
16) $21x^2 + 29x - 10$

17) $(5x + 2y)$
18) $16x^2 - 2x - \frac{15}{2}$
19) $36x^2 + 24x + 4$
20) $40x$
21) $x^3 + 6x^2 + 12x + 8$
22) $(9x + y)$

ATI TEAS 6 Subject Test – Mathematics

# Chapter 9 : Functions Operations and Quadratic

**Topics that you'll practice in this chapter:**

- ✓ Evaluating Function
- ✓ Adding and Subtracting Functions
- ✓ Multiplying and Dividing Functions
- ✓ Composition of Functions
- ✓ Quadratic Equation
- ✓ Solving Quadratic Equations
- ✓ Quadratic Formula and the Discriminant
- ✓ Graphing Quadratic Functions

*It's fine to work on any problem, so long as it generates interesting mathematics along the way – even if you don't solve it at the end of the day." – Andrew Wiles*

# ATI TEAS 6 Subject Test – Mathematics

## Evaluating Function

✍ **Write each of following in function notation.**

1) $h = -8x + 3$

2) $k = 2a - 14$

3) $d = 11t$

4) $y = \frac{5}{12}x - \frac{7}{12}$

5) $m = 24n - 210$

6) $c = p^2 - 5p + 10$

✍ **Evaluate each function.**

7) $f(x) = 2x - 7$, find $f(-3)$

8) $g(x) = \frac{1}{9}x + 12$, find $f(18)$

9) $h(x) = -4x + 9$, find $f(3)$

10) $f(x) = -x + 19$, find $f(-3)$

11) $f(a) = 7a - 12$, find $f(3)$

12) $h(x) = 14 - 3x$, find $f(-4)$

13) $g(n) = 6n - 10$, find $f(2)$

14) $f(x) = -11x - 4$, find $f(-1)$

15) $k(n) = -20 - 3.5n$, find $f(2)$

16) $f(x) = -0.7x + 3.3$, find $f(-7)$

17) $g(n) = \frac{11n+8}{n}$, find $g(2)$

18) $g(n) = \sqrt{3n} + 12$, find $g(3)$

19) $h(x) = x^{-2} - 7$, find $h(\frac{1}{9})$

20) $h(n) = n^{-3} + 11$, find $h(\frac{1}{4})$

21) $h(n) = n^3 - 2$, find $h(\frac{1}{2})$

22) $h(n) = n^2 - 4$, find $h(-\frac{1}{3})$

23) $h(n) = 4n^2 - 13$, find $h(-5)$

24) $h(n) = -2n^3 - 6n$, find $h(2)$

25) $g(n) = \sqrt{16n^2} - \sqrt{n}$, find $g(4)$

26) $h(a) = \frac{-14a+9}{3a}$, find $h(-b)$

27) $k(a) = 12a - 14$, find $k(a - 3)$

28) $h(x) = \frac{1}{9}x + 18$, find $h(-18x)$

29) $h(x) = 8x^2 + 16$, find $h(\frac{x}{2})$

30) $h(x) = x^4 - 20$, find $h(-2x)$

ATI TEAS 6 Subject Test – Mathematics

# Adding and Subtracting Functions

✏️ **Perform the indicated operation.**

1) $f(x) = 2x + 3$

   $g(x) = x + 7$

   Find $(f - g)(2)$

2) $g(a) = -5a - 8$

   $f(a) = -3a - 5$

   Find $(g - f)(-2)$

3) $h(t) = 4t + 3$

   $g(t) = 4t + 7$

   Find $(h - g)(t)$

4) $g(a) = -6a - 10$

   $f(a) = 3a^2 + 9$

   Find $(g - f)(x)$

5) $g(x) = \frac{5}{6}x - 23$

   $h(x) = \frac{5}{12}x + 25$

   Find $g(12) - h(12)$

6) $h(x) = \sqrt{3x} - 2$

   $g(x) = \sqrt{3x} + 5$

   Find $(h + g)(12)$

7) $f(x) = x^{-1}$

   $g(x) = x^2 + \frac{5}{x}$

   Find $(f - g)(-3)$

8) $h(n) = n^2 + 2$

   $g(n) = -4n + 6$

   Find $(h - g)(2a)$

9) $g(x) = -2x^2 - 5 - 4x$

   $f(x) = 7 + 2x$

   Find $(g - f)(3x)$

10) $g(t) = 11t - 4$

    $f(t) = -2t^2 + 5$

    Find $(g + f)(-t)$

11) $f(x) = 8x + 9$

    $g(x) = -5x^2 + 3x$

    Find $(f - g)(-x^2)$

12) $f(x) = -3x^4 - 5x$

    $g(x) = 2x^4 + 5x + 22$

    Find $(f + g)(3x^2)$

WWW.MathNotion.Com

ATI TEAS 6 Subject Test – Mathematics

## Multiplying and Dividing Functions

✎ **Perform the indicated operation.**

1) $g(x) = -2x - 1$
   $f(x) = 4x + 3$
   Find $(g.f)(2)$

2) $f(x) = 5x$
   $h(x) = -2x + 3$
   Find $(f.h)(-2)$

3) $g(a) = 5a - 2$
   $h(a) = 2a - 3$
   Find $(g.h)(-3)$

4) $f(x) = 2x - 7$
   $h(x) = x - 5$
   Find $\left(\frac{f}{h}\right)(4)$

5) $f(x) = 8a^2$
   $g(x) = 3 + 2a$
   Find $\left(\frac{f}{g}\right)(2)$

6) $g(a) = \sqrt{4a} + 2$
   $f(a) = (-a)^4 + 1$
   Find $\left(\frac{g}{f}\right)(1)$

7) $g(t) = t^3 + 1$
   $h(t) = 5t - 2$
   Find $(g.h)(-2)$

8) $g(n) = n^2 + 2n - 4$
   $h(n) = -5n + 3$
   Find $(g.h)(1)$

9) $g(a) = (a - 3)^2$
   $f(a) = a^2 + 4$
   Find $\left(\frac{g}{f}\right)(3)$

10) $g(x) = -3x^2 + \frac{4}{5}x + 9$
    $f(x) = x^2 - 24$
    Find $\left(\frac{g}{f}\right)(5)$

11) $f(x) = 2x^3 - 5x^2 + 1$
    $g(x) = 3x - 1$
    Find $(f.g)(x)$

12) $f(x) = 5x - 2$
    $g(x) = x^3 - 2x$
    Find $(f.g)(x^2)$

WWW.MathNotion.Com

# ATI TEAS 6 Subject Test – Mathematics

## Composition of Functions

✎ Using $f(x) = 2x - 5$ and $g(x) = -2x$, find:

1) $f(g(2)) =$

2) $f(g(-1)) =$

3) $g(f(-4)) =$

4) $g(f(5)) =$

5) $f(g(3)) =$

6) $g(f(0)) =$

✎ Using $f(x) = -\frac{1}{4}x + \frac{3}{4}$ and $g(x) = 2x^2$, find:

7) $g(f(-2)) =$

8) $g(f(4)) =$

9) $g(g(1)) =$

10) $f(f(1)) =$

11) $g(f(-4)) =$

12) $f(g(x)) =$

✎ Using $f(x) = -2x + 2$ and $g(x) = x + 1$, find:

13) $g(f(1)) =$

14) $f(f(0)) =$

15) $f(g(-1)) =$

16) $f(g(-3)) =$

17) $g(f(2)) =$

18) $f(g(x)) =$

✎ Using $f(x) = \sqrt{x + 9}$ and $g(x) = x - 9$, find:

19) $f(g(9)) =$

20) $g(f(-9)) =$

21) $f(g(4)) =$

22) $f(f(7)) =$

23) $g(f(-5)) =$

24) $g(g(0)) =$

ATI TEAS 6 Subject Test – Mathematics

# Quadratic Equation

### ✏ Multiply.

1) $(x-4)(x+6) = $ _____

2) $(x+5)(x+7) = $ _____

3) $(x-6)(x+8) = $ _____

4) $(x+2)(x-9) = $ _____

5) $(x-7)(x-8) = $ _____

6) $(3x+2)(x-3) = $ _____

7) $(4x-3)(x+2) = $ _____

8) $(4x-5)(x+1) = $ _____

9) $(7x+1)(x-6) = $ _____

10) $(5x+1)(3x-3) = $ _____

### ✏ Factor each expression.

11) $x^2 - 2x - 8 = $ _____

12) $x^2 + 8x + 15 = $ _____

13) $x^2 - 2x - 24 = $ _____

14) $x^2 - 10x + 21 = $ _____

15) $x^2 + 10x + 21 = $ _____

16) $4x^2 + 9x + 5 = $ _____

17) $5x^2 + 13x - 6 = $ _____

18) $5x^2 + 17x - 12 = $ _____

19) $2x^2 + 7x + 5 = $ _____

20) $9x^2 - 21x + 6 = $ _____

### ✏ Calculate each equation.

21) $(x+6)(x-3) = 0$

22) $(x+1)(x+8) = 0$

23) $(3x+6)(x+5) = 0$

24) $(2x-2)(4x+8) = 0$

25) $x^2 + x + 10 = 22$

26) $x^2 + 11x + 36 = 12$

27) $2x^2 + 9x + 9 = 5$

28) $x^2 + 3x - 24 = 4$

29) $5x^2 + 5x - 40 = 20$

30) $8x^2 + 8x = 48$

ATI TEAS 6 Subject Test – Mathematics

## Solving Quadratic Equations

✎ **Solve each equation by factoring or using the quadratic formula.**

1) $(x+9)(x-1) = 0$

2) $(x+7)(x+6) = 0$

3) $(x-8)(x+3) = 0$

4) $(x-6)(x-4) = 0$

5) $(x+2)(x+12) = 0$

6) $(5x+4)(x+7) = 0$

7) $(6x+1)(4x+5) = 0$

8) $(2x+7)(x+8) = 0$

9) $(x+6)(3x+15) = 0$

10) $(12x+2)(x+8) = 0$

11) $x^2 = 8x$

12) $x^2 - 16 = 0$

13) $3x^2 + 6 = 9x$

14) $-2x^2 - 8 = 10x$

15) $5x^2 + 40x = 45$

16) $x^2 + 10x = 24$

17) $x^2 + 6x = 16$

18) $x^2 + 9x = -18$

19) $x^2 + 13x = -36$

20) $x^2 + 3x - 15 = 5x$

21) $x^2 + 8x + 7 = -8$

22) $3x^2 - 11x = -9 + x$

23) $10x^2 + 3 = 27x - 15$

24) $7x^2 - 6x + 8 = 8$

25) $2x^2 - 12 = -3x + 2$

26) $10x^2 - 26x - 3 = -15$

27) $3x^2 + 21 = -16x + 5$

28) $x^2 + 15x - 10 = -66$

29) $3x^2 - 8x - 8 = 4 + x$

30) $2x^2 + 6x - 24 = 12$

31) $3x^2 - 33x + 54 = -18$

32) $-10x^2 - 15x - 9 = -9 - 27x^2$

WWW.MathNotion.Com

# ATI TEAS 6 Subject Test – Mathematics

## Quadratic Formula and the Discriminant

✎ Find the value of the discriminant of each quadratic equation.

1) $3x(x-8) = 0$

2) $2x^2 + 6x - 4 = 0$

3) $x^2 + 6x + 7 = 0$

4) $x^2 - x + 3 = 0$

5) $x^2 + 4x - 3 = 0$

6) $2x^2 + 6x - 10 = 0$

7) $3x^2 + 7x + 5 = 0$

8) $x^2 - 6x - 4 = 0$

9) $2x^2 + 8x + 3 = 0$

10) $x^2 + 7x - 5 = 0$

11) $5x^2 + 2x - 3 = 0$

12) $-3x^2 - 11x + 4 = 0$

13) $-6x^2 - 12x + 8 = 0$

14) $-x^2 - 9x - 12 = 0$

15) $7x^2 - 6x - 10 = 0$

16) $-4x^2 - 2x + 8 = 0$

17) $5x^2 + 8x - 2 = 0$

18) $6x^2 - 4x = 0$

19) $3x^2 - 5x + 2 = 0$

20) $4x^2 + 9x + 3 = 0$

✎ Find the discriminant of each quadratic equation then state the number of real and imaginary solutions.

21) $-4x^2 - 16 = 16x$

22) $20x^2 = 20x - 5$

23) $-11x^2 - 19x = 26$

24) $22x^2 - 4x + 1 = 18x^2$

25) $-11x^2 = -15x + 8$

26) $3x^2 + 6x + 9 = 6$

27) $13x^2 - 5x - 12 = -26$

28) $-8x^2 - 32x - 25 = 7$

WWW.MathNotion.Com

ATI TEAS 6 Subject Test – Mathematics

# Graphing Quadratic Functions

✎ Sketch the graph of each function. Identify the vertex and axis of symmetry.

1) $y = (x + 3)^2 + 2$

2) $y = (x - 3)^2 - 2$

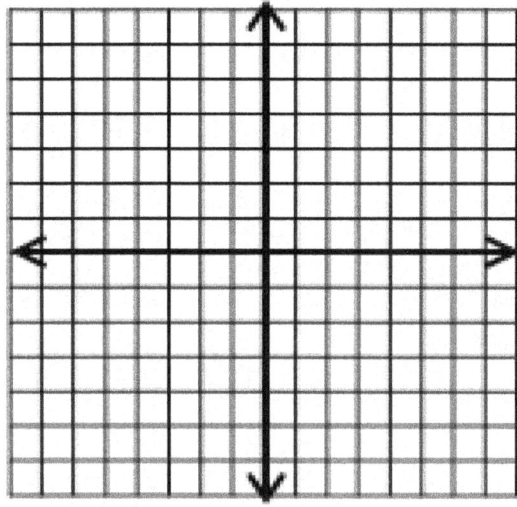

3) $y = 6 - (-x + 4)^2$

4) $y = -3x^2 - 6x + 9$

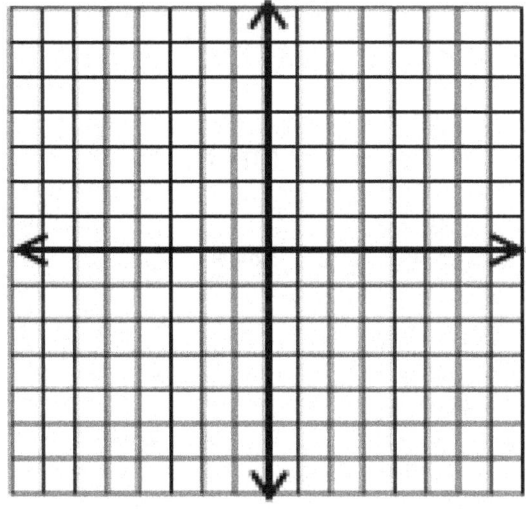
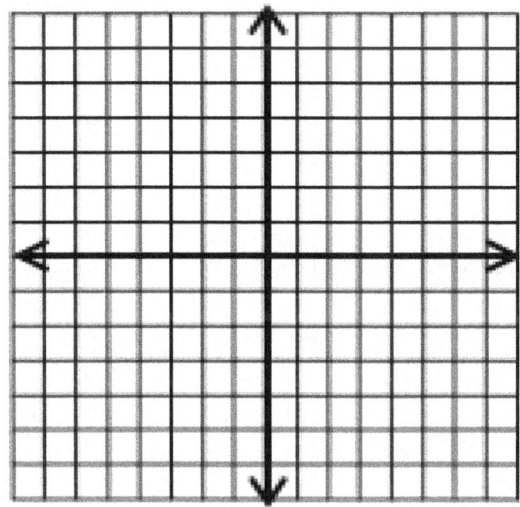

WWW.MathNotion.Com

# ATI TEAS 6 Subject Test – Mathematics

## Answers of Worksheets

**Evaluating Function**

1) $h(x) = -8x + 3$
2) $k(a) = 2a - 14$
3) $d(t) = 11t$
4) $f(x) = \frac{5}{12}x - \frac{7}{12}$
5) $m(n) = 24n - 210$
6) $c(p) = p^2 - 5p + 10$
7) $-13$
8) $14$
9) $-3$
10) $22$
11) $9$
12) $26$
13) $2$
14) $7$
15) $-27$
16) $8.2$
17) $15$
18) $15$
19) $74$
20) $75$
21) $-1\frac{7}{8}$
22) $-3\frac{8}{9}$
23) $87$
24) $-28$
25) $14$
26) $-\frac{14b+9}{3b}$
27) $12a - 50$
28) $-2x + 18$
29) $2x^2 + 16$
30) $16x^4 - 20$

**Adding and Subtracting Functions**

1) $-2$
2) $1$
3) $-4$
4) $-3x^2 - 6x - 19$
5) $-43$
6) $15$
7) $-7\frac{2}{3}$
8) $4a^2 + 8a - 4$
9) $-18x^2 - 18x - 12$
10) $-2t^2 - 11t + 1$
11) $5x^4 - 5x^2 + 9$
12) $-81x^8 + 22$

**Multiplying and Dividing Functions**

1) $-55$
2) $-70$
3) $153$
4) $-1$
5) $4\frac{4}{7}$
6) $2$
7) $84$
8) $2$
9) $0$
10) $-62$
11) $6x^4 - 17x^3 + 5x^2 + 3x - 1$
12) $5x^8 - 2x^6 - 10x^4 + 4x^2$

**Composition of Functions**

1) $-13$
2) $-1$
3) $26$
4) $-10$
5) $-17$
6) $10$
7) $\frac{25}{8}$
8) $\frac{1}{8}$
9) $8$
10) $\frac{5}{8}$
11) $\frac{49}{8}$
12) $-\frac{1}{2}(x^2 - \frac{3}{2})$
13) $1$
14) $-2$
15) $2$
16) $6$
17) $-1$
18) $-2x$

# ATI TEAS 6 Subject Test – Mathematics

19) 3
20) −9
21) 2
22) $\sqrt{13}$
23) −7
24) −18

## Quadratic Equations

1) $x^2 + 2x - 24$
2) $x^2 + 12x + 35$
3) $x^2 + 2x - 48$
4) $x^2 - 7x - 18$
5) $x^2 - 15x + 56$
6) $3x^2 - 7x - 6$
7) $4x^2 + 5x - 6$
8) $4x^2 - x - 5$
9) $7x^2 - 41x - 6$
10) $15x^2 - 12x - 3$
11) $(x - 4)(x + 2)$
12) $(x + 5)(x + 3)$
13) $(x - 6)(x + 4)$
14) $(x - 3)(x - 7)$
15) $(x + 3)(x + 7)$
16) $(4x + 5)(x + 1)$
17) $(5x - 2)(x + 3)$
18) $(5x - 3)(x + 4)$
19) $(2x + 5)(x + 1)$
20) $3(x - 2)(3x - 1)$
21) $x = -6, x = 3$
22) $x = -1, x = -8$
23) $x = -2, x = -5$
24) $x = 1, x = -2$
25) $x = 3, x = -4$
26) $x = -3, x = -8$
27) $x = -4, x = -\frac{1}{2}$
28) $x = 4, x = -7$
29) $x = 3, x = -4$
30) $x = -3, x = 2$

## Solving quadratic equations

1) $\{-9, 1\}$
2) $\{-6, -7\}$
3) $\{8, -3\}$
4) $\{6, 4\}$
5) $\{-2, -12\}$
6) $\{-\frac{4}{5}, -7\}$
7) $\{-\frac{5}{4}, -\frac{1}{6}\}$
8) $\{-\frac{7}{2}, -8\}$
9) $\{-6, -5\}$
10) $\{-\frac{1}{6}, -8\}$
11) $\{8, 0\}$
12) $\{4, -4\}$
13) $\{2, 1\}$
14) $\{-4, -1\}$
15) $\{1, -9\}$
16) $\{2, -12\}$
17) $\{2, -8\}$
18) $\{-3, -6\}$
19) $\{-4, -9\}$
20) $\{5, -3\}$
21) $\{-5, -3\}$
22) $\{1, 3\}$
23) $\{\frac{6}{5}, \frac{3}{2}\}$
24) $\{\frac{6}{7}, 0\}$
25) $\{-\frac{7}{2}, 2\}$
26) $\{\frac{3}{5}, 2\}$
27) $\{-\frac{4}{3}, -4\}$
28) $\{-8, -7\}$
29) $\{4, -1\}$
30) $\{3, -6\}$
31) $\{3, 8\}$
32) $\{\frac{15}{17}, 0\}$

## Quadratic formula and the discriminant

1) 576
2) 68
3) 8
4) −11
5) 28
6) 116
7) −11
8) 52
9) 40
10) 69
11) 64
12) 169
13) 336
14) 33
15) 316
16) 132
17) 104
18) 16
19) 1
20) 33
21) 0, one real solution
22) 0, one real solution
23) −783, no solution

# ATI TEAS 6 Subject Test – Mathematics

24) 0, *one real solution*

25) −127, *no solution*

26) 0, *one real solution*

27) −703, *no solution*

28) 0, *one real solution*

**Graphing quadratic functions**

1) $(-3, 2), x = -3$

2) $(3, -2), x = 3$

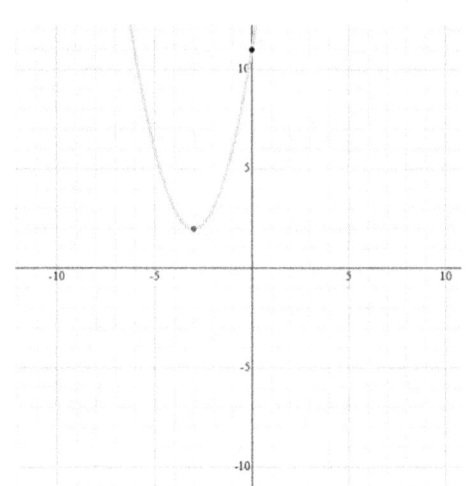
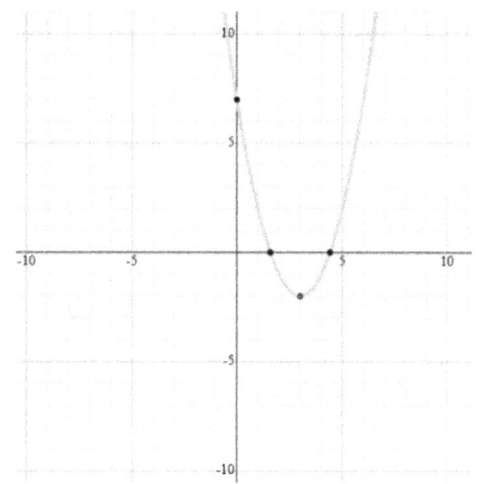

3) $(4, 6), x = 4$

4) $(-1, 12), x = -1$

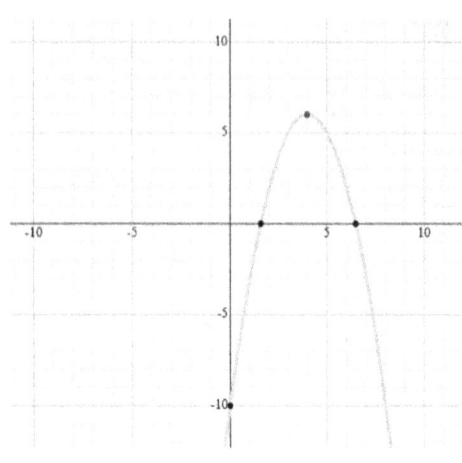
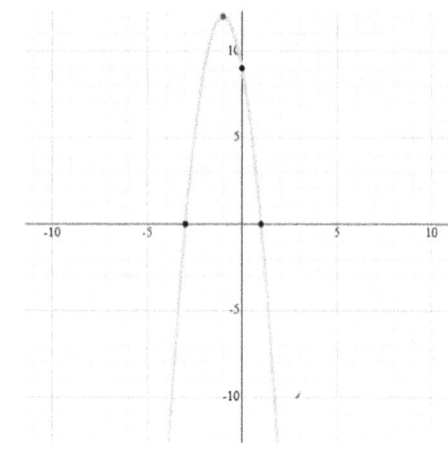

ATI TEAS 6 Subject Test – Mathematics

# Chapter 10 : Geometry and Solid Figures

**Topics that you'll practice in this chapter:**

- ✓ Angles
- ✓ Pythagorean Relationship
- ✓ Triangles
- ✓ Polygons
- ✓ Trapezoids
- ✓ Circles
- ✓ Cubes
- ✓ Rectangular Prism
- ✓ Cylinder
- ✓ Pyramids and Cone

*Mathematics is, as it were, a sensuous logic, and relates to philosophy as do the arts, music, and plastic art to poetry.* — K. Shegel

# ATI TEAS 6 Subject Test – Mathematics

## Angles

✎ **What is the value of *x* in the following figures?**

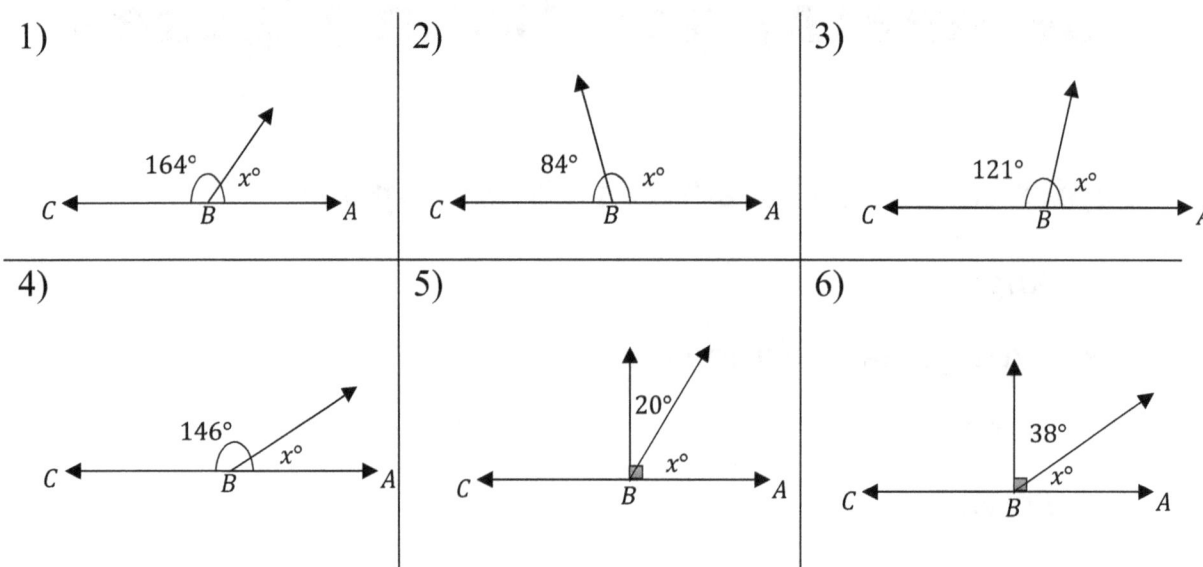

✎ **Calculate.**

7) Two supplement angles have equal measures. What is the measure of each angle? _____

8) The measure of an angle is seven fifth the measure of its supplement. What is the measure of the angle? _____

9) Two angles are complementary and the measure of one angle is 24 less than the other. What is the measure of the smaller angle? _____

10) Two angles are complementary. The measure of one angle is one fifth the measure of the other. What is the measure of the bigger angle? _____

11) Two supplementary angles are given. The measure of one angle is 40° less than the measure of the other. What does the smaller angle measure? _____

WWW.MathNotion.Com

ATI TEAS 6 Subject Test – Mathematics

# Pythagorean Relationship

 Do the following lengths form a right triangle?

1)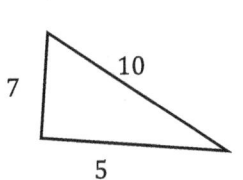

2)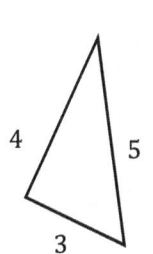

3)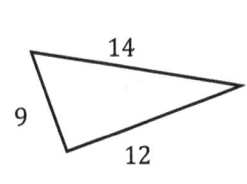

4)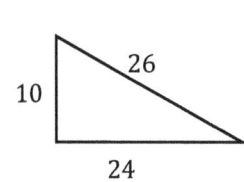

5)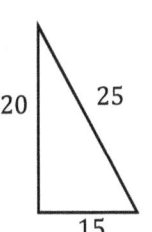

6)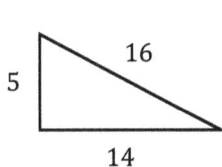

7)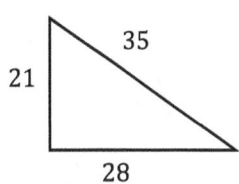

8)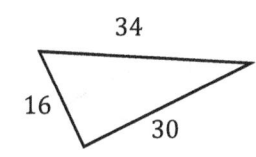

 Find the missing side?

9)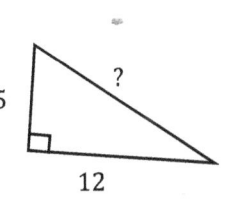

10)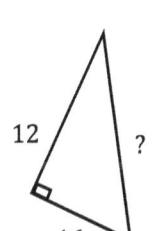

11)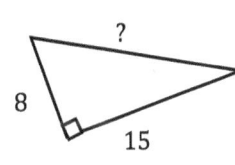

12)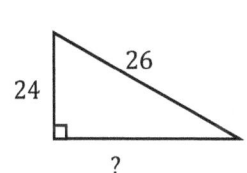

13)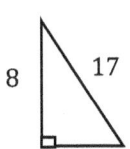

14)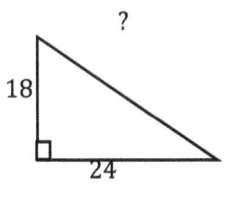

15)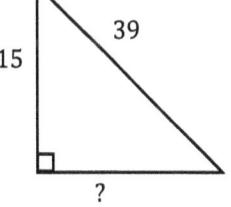

16)

WWW.MathNotion.Com 125

ATI TEAS 6 Subject Test – Mathematics

# Triangles

✎ Find the measure of the unknown angle in each triangle.

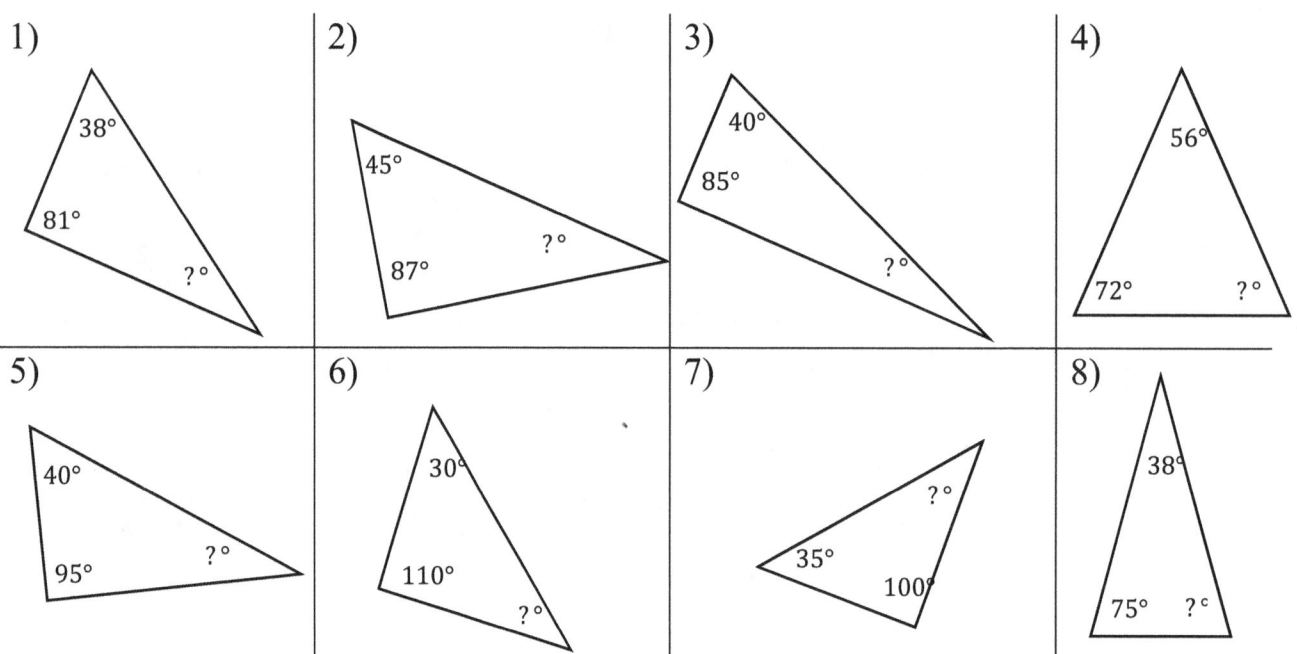

✎ Find area of each triangle.

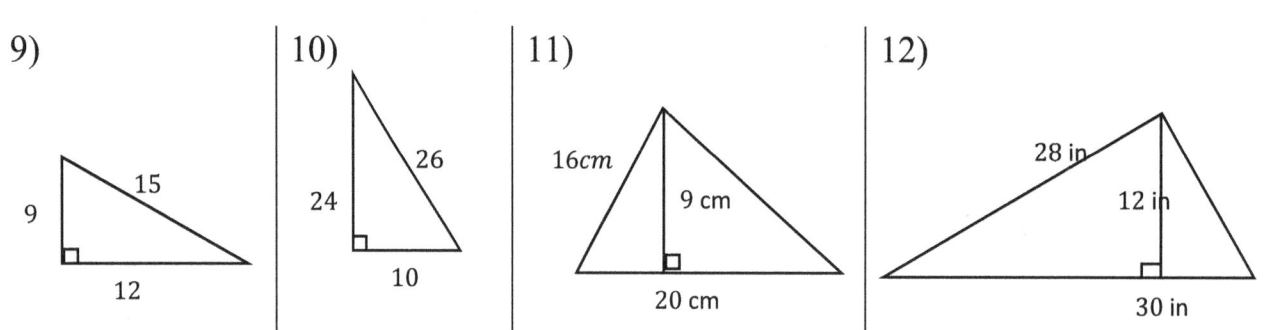

# ATI TEAS 6 Subject Test – Mathematics

# Polygons

✏️ **Find the perimeter of each shape.**

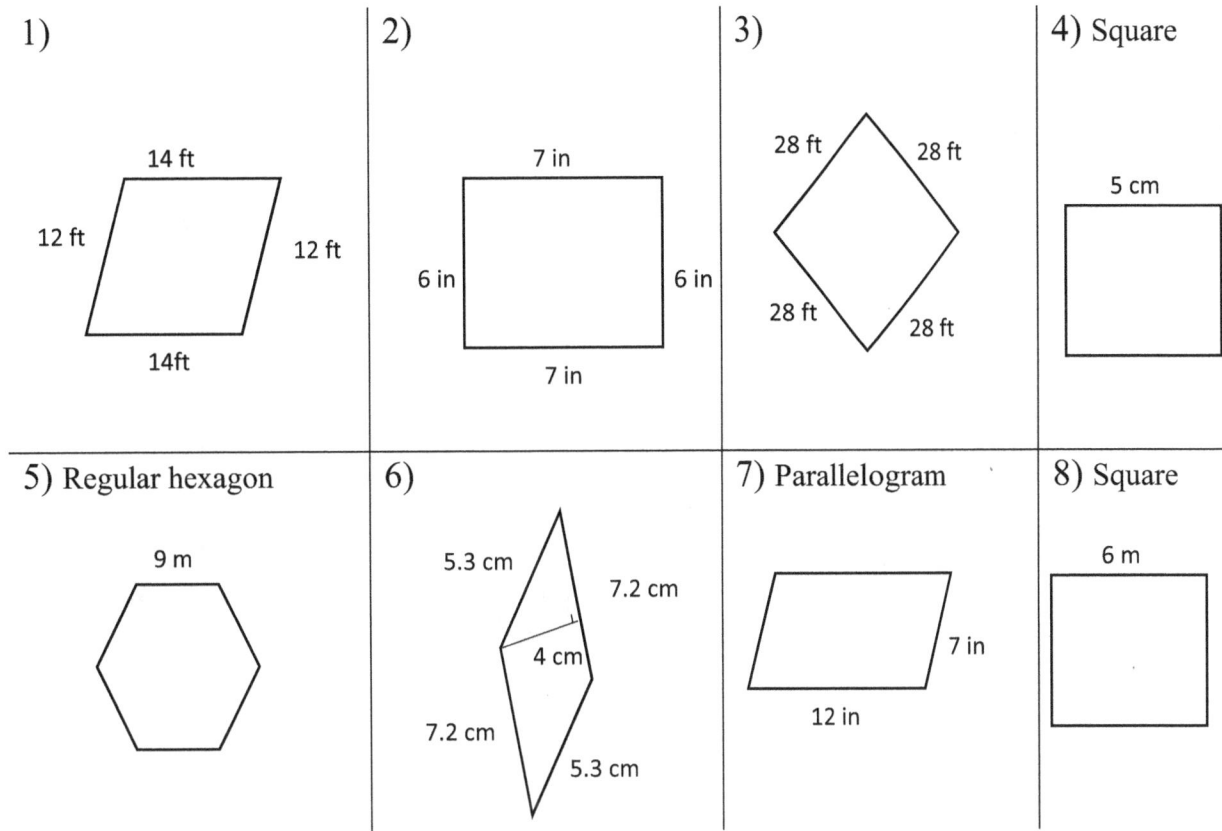

✏️ **Find the area of each shape.**

# ATI TEAS 6 Subject Test – Mathematics

## Trapezoids

✏️ **Find the area of each trapezoid.**

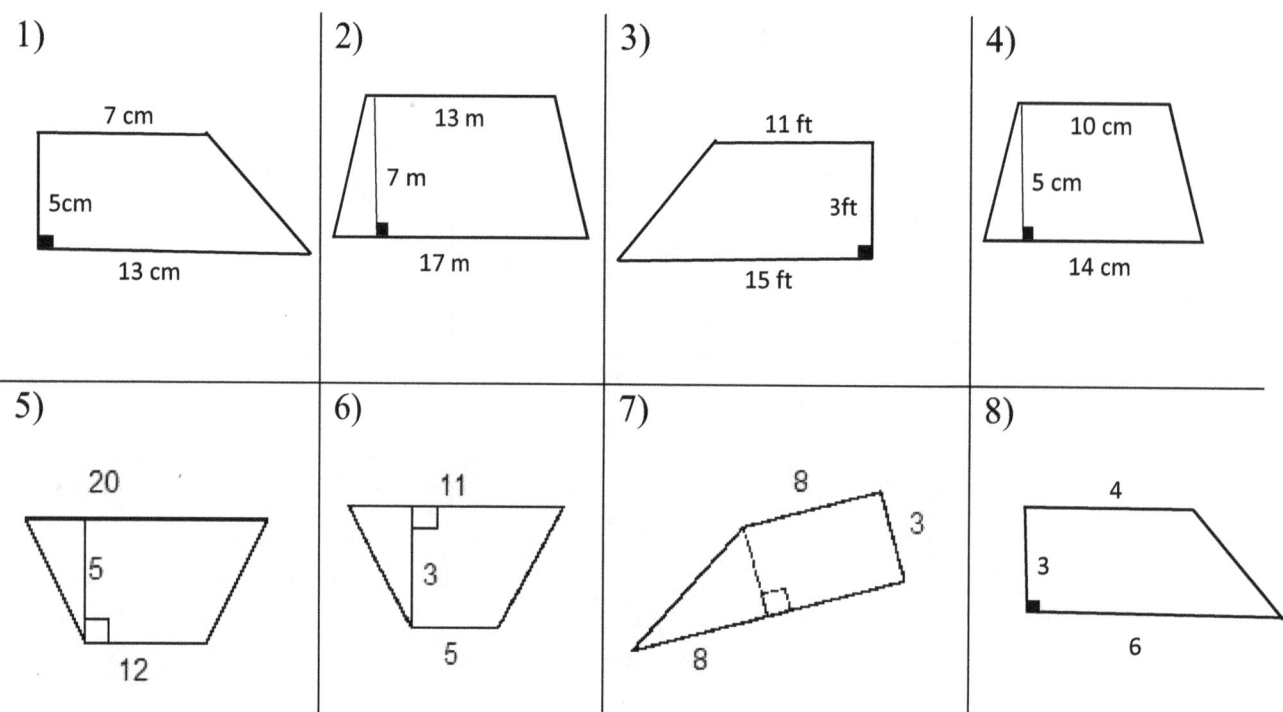

✏️ **Calculate.**

1) A trapezoid has an area of 45 cm² and its height is 5 cm and one base is 5 cm. What is the other base length? _____

2) If a trapezoid has an area of 99 ft² and the lengths of the bases are 8 ft and 10 ft, find the height? _____

3) If a trapezoid has an area of 126 m² and its height is 14 m and one base is 6 m, find the other base length? _____

4) The area of a trapezoid is 440 ft² and its height is 22 ft. If one base of the trapezoid is 15 ft, what is the other base length? _____

WWW.MathNotion.Com

# ATI TEAS 6 Subject Test – Mathematics

## Circles

✏️ **Find the area of each circle.** ($\pi = 3.14$)

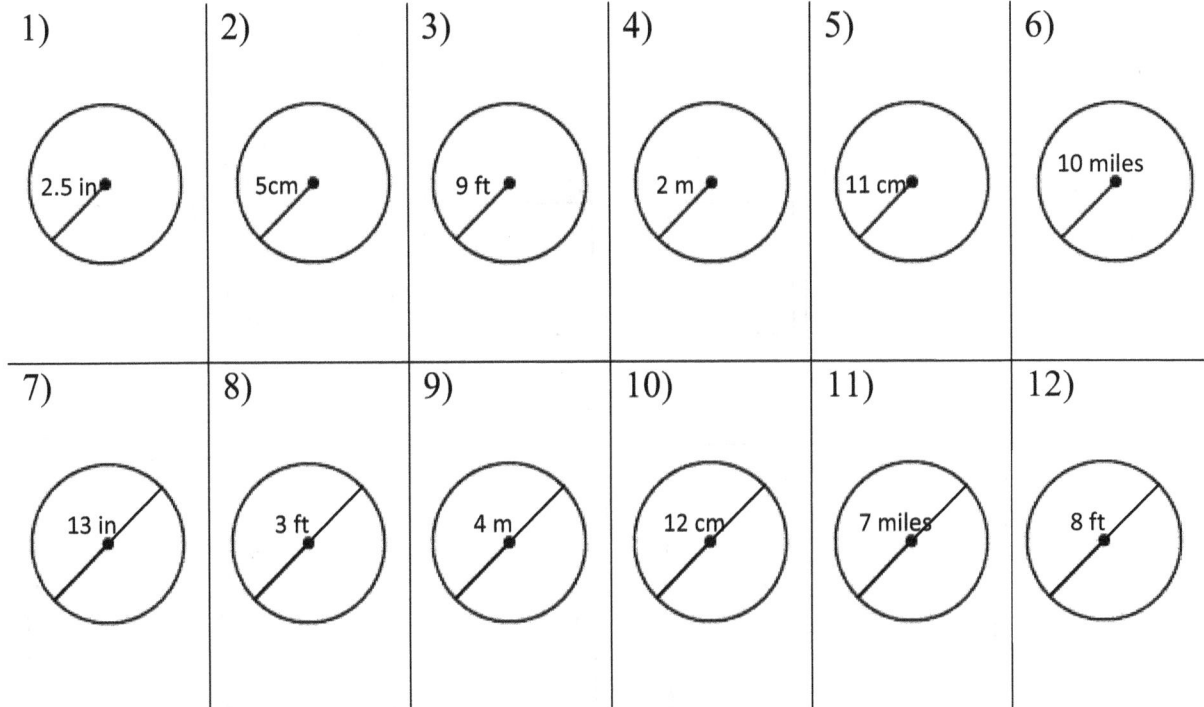

✏️ **Complete the table below.** ($\pi = 3.14$)

| Circle No. | Radius | Diameter | Circumference | Area |
|---|---|---|---|---|
| 1 | 1 in | 2 in | 6.28 in | 3.14 $in^2$ |
| 2 |  | 10 m |  |  |
| 3 |  |  |  | 28.26 $ft^2$ |
| 4 |  |  | 47.1 mi |  |
| 5 |  | 11 km |  |  |
| 6 | 7 cm |  |  |  |
| 7 |  | 12 ft |  |  |
| 8 |  |  |  | 314 $m^2$ |
| 9 |  |  | 56.52 in |  |
| 10 | 4.5 ft |  |  |  |

# ATI TEAS 6 Subject Test – Mathematics

## Cubes

✎ Find the volume of each cube.

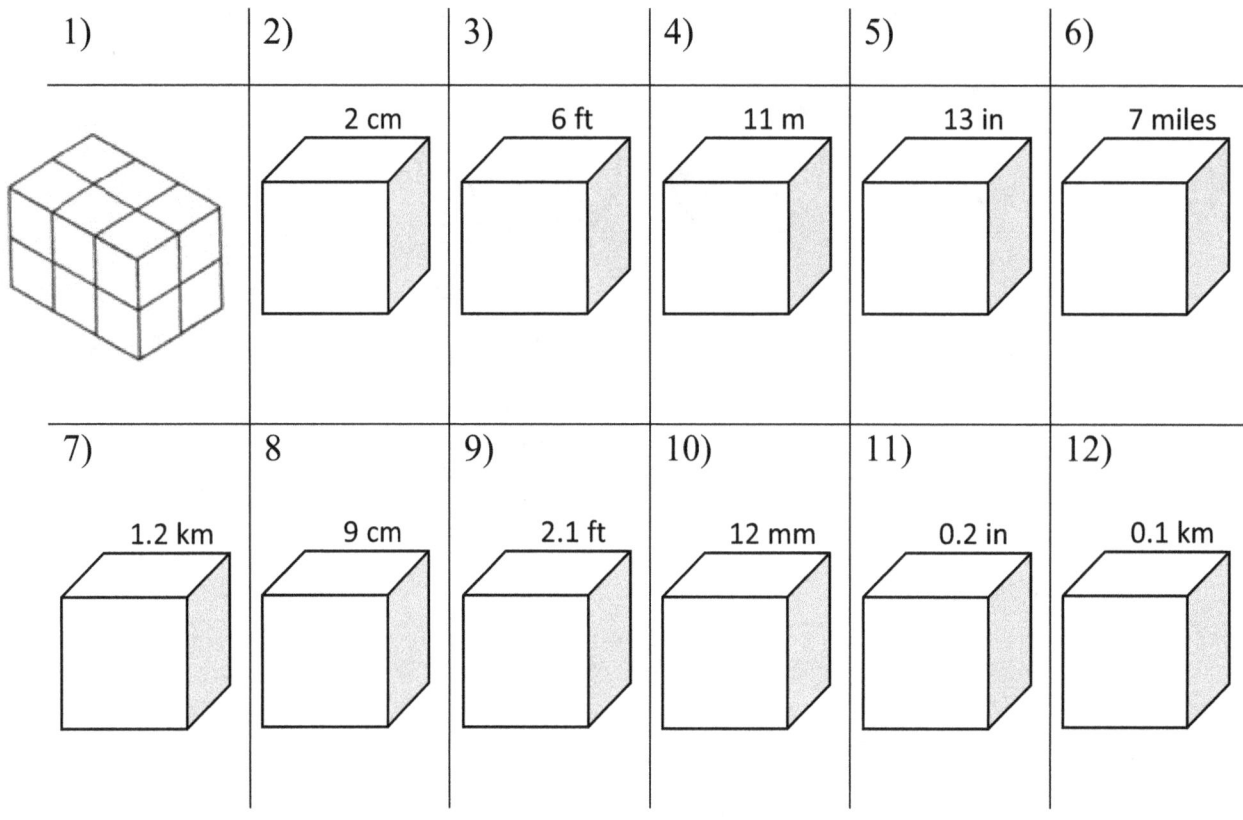

✎ Find the surface area of each cube.

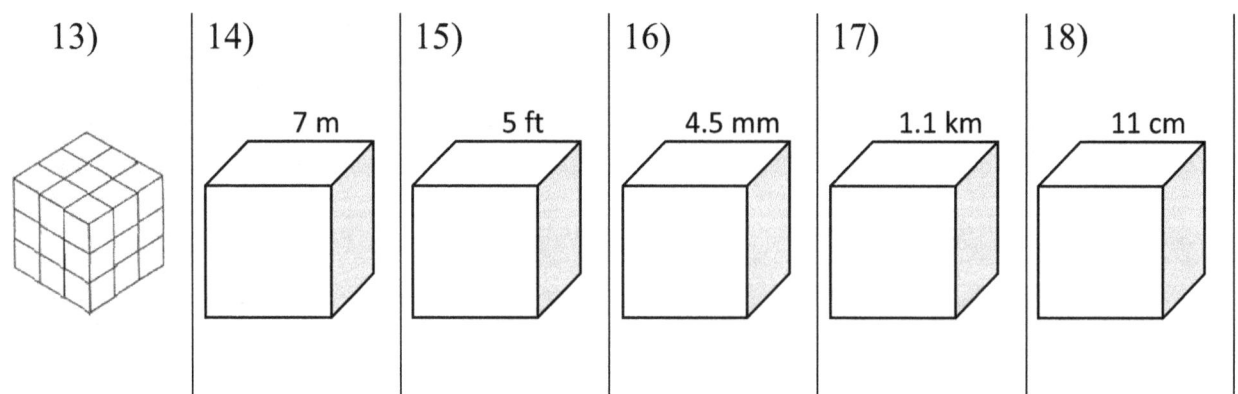

ATI TEAS 6 Subject Test – Mathematics

# Rectangular Prism

✎ Find the volume of each Rectangular Prism.

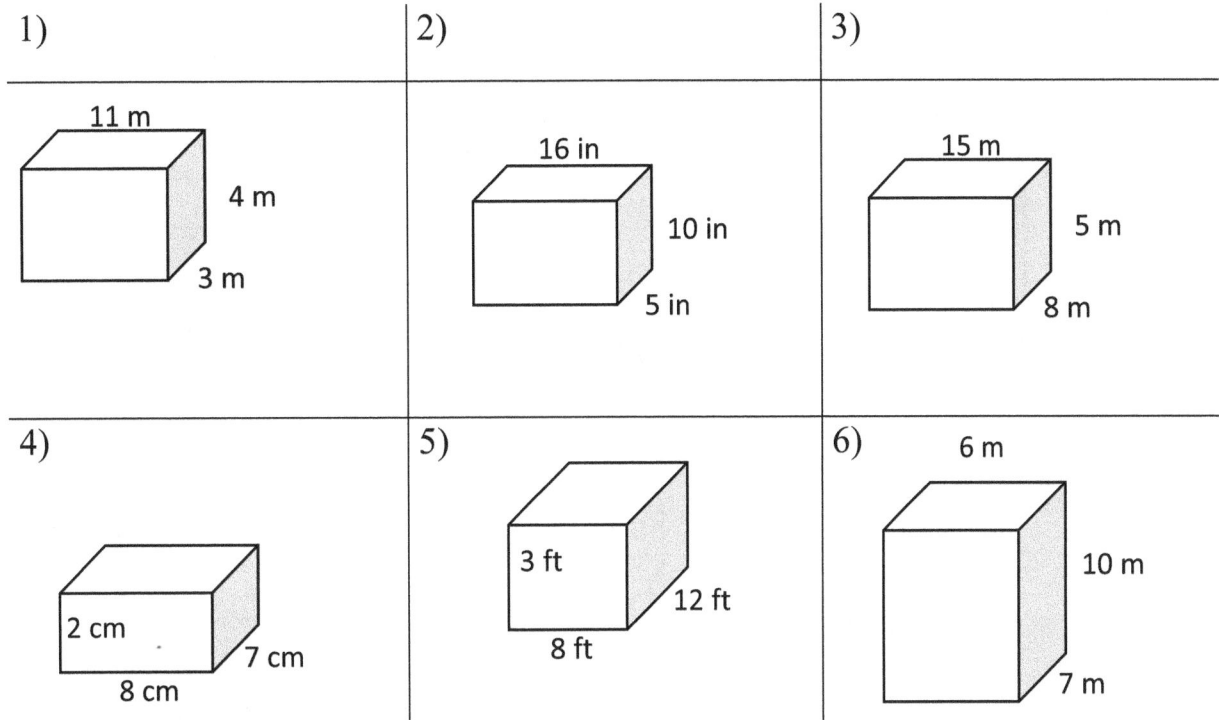

✎ Find the surface area of each Rectangular Prism.

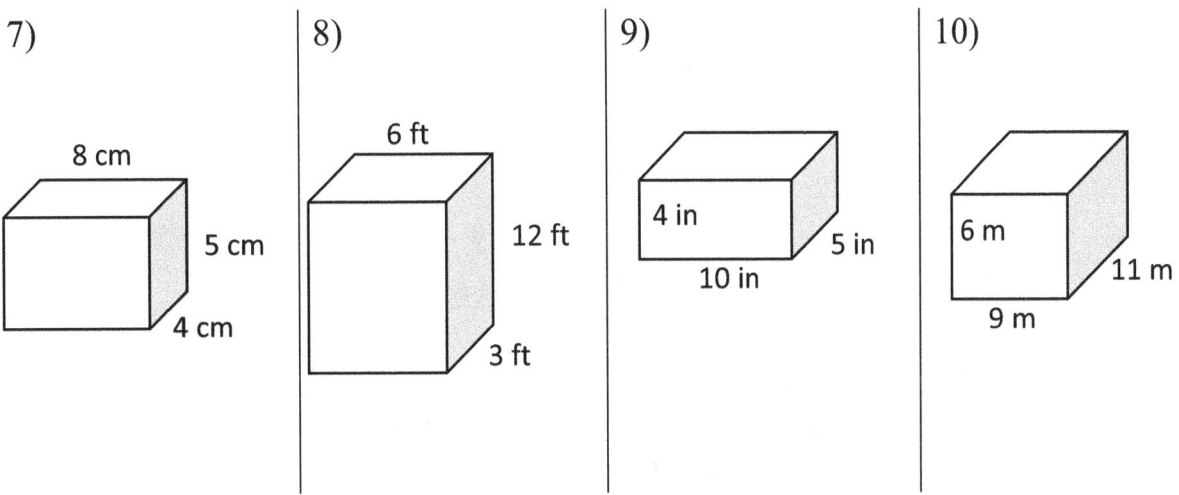

WWW.MathNotion.Com

# ATI TEAS 6 Subject Test – Mathematics

## Cylinder

✎ **Find the volume of each Cylinder. Round your answer to the nearest tenth.** ($\pi = 3.14$)

1)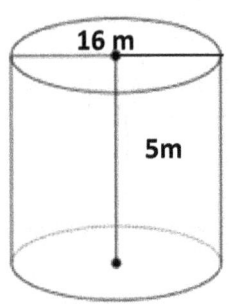

2)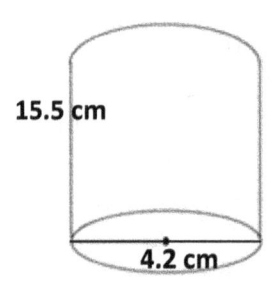

3)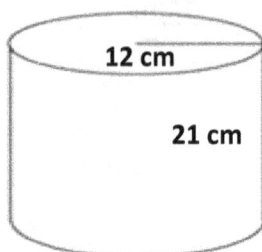

4)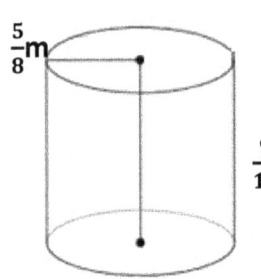

5)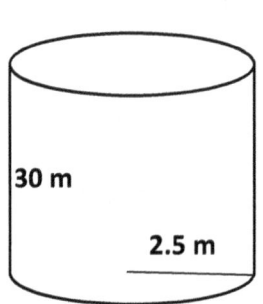

6)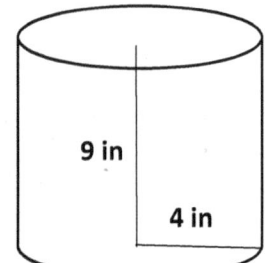

✎ **Find the surface area of each Cylinder.** ($\pi = 3.14$)

7)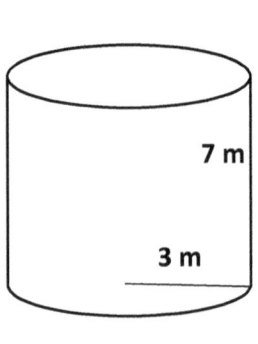

8)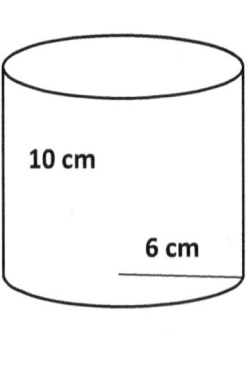

9)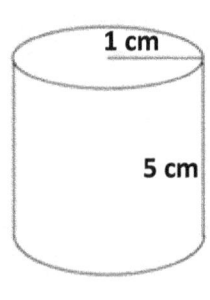

10)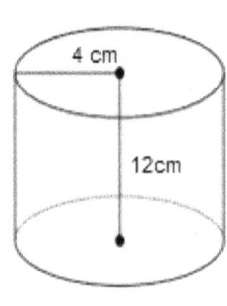

WWW.MathNotion.Com

ATI TEAS 6 Subject Test – Mathematics

# Pyramids and Cone

✎ Find the volume of each Pyramid and Cone. ($\pi = 3.14$)

1)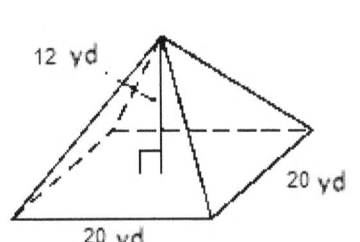

2)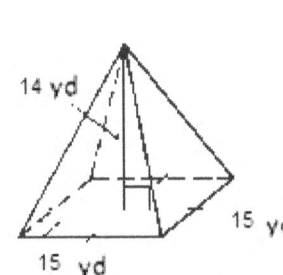

3)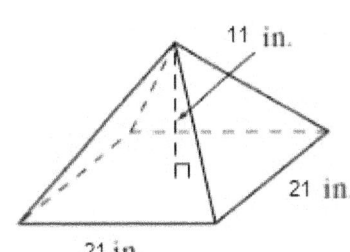

4)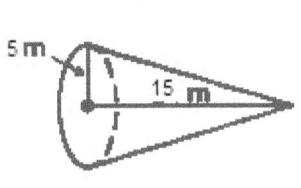

5)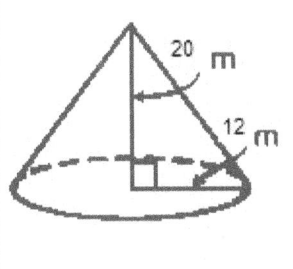

6)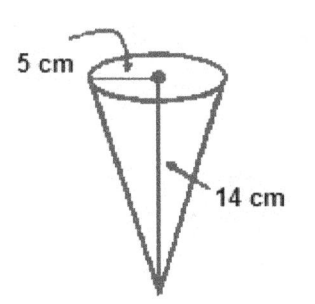

✎ Find the surface area of each Pyramid and Cone. ($\pi = 3.14$)

7)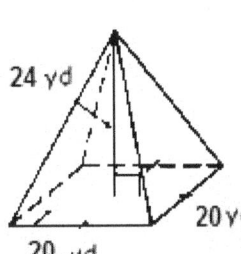

8)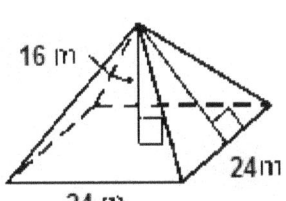

9)

10)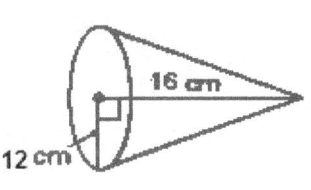

WWW.MathNotion.Com

# ATI TEAS 6 Subject Test – Mathematics

## Answers of Worksheets

### Angles
1) 16°
2) 96°
3) 59°
4) 34°
5) 70°
6) 52°
7) 90°
8) 75°
9) 33°
10) 75°
11) 70°

### Pythagorean Relationship
1) No
2) Yes
3) No
4) Yes
5) Yes
6) No
7) Yes
8) Yes
9) 13
10) 20
11) 17
12) 10
13) 15
14) 30
15) 36
16) 12

### Triangles
1) 60°
2) 48°
3) 55°
4) 52°
5) 45°
6) 40°
7) 45°
8) 67°
9) 54 square unites
10) 120 square unites
11) 90 square unites
12) 180 square unites

### Polygons
1) 52 ft
2) 26 in
3) 112 ft
4) 20 cm
5) 54 m
6) 25 cm
7) 38 in
8) 24 m
9) 30 $m^2$
10) 300 $in^2$
11) 160 $km^2$
12) 49 $in^2$

### Trapezoids
1) 50 $cm^2$
2) 105 $m^2$
3) 39 $ft^2$
4) 60 $cm^2$
5) 80
6) 24
7) 36
8) 15

### Calculate
1) 13 cm
2) 11 ft
3) 12 m
4) 25 ft

### Circles
1) 19.63 $in^2$
2) 78.5 $cm^2$
3) 254.34 $ft^2$
4) 12.56 $m^2$
5) 379.94 $cm^2$
6) 314 $miles^2$
7) 132.67 $in^2$
8) 7.07 $ft^2$
9) 12.56 $m^2$
10) 113.04 $cm^2$
11) 38.47 $miles^2$
12) 50.24 $ft^2$

# ATI TEAS 6 Subject Test – Mathematics

| Circle No. | Radius | Diameter | Circumference | Area |
|---|---|---|---|---|
| 1 | 1 in | 2 in | 6.28 in | 3.14 $in^2$ |
| 2 | 5 m | 10 m | 31.4 m | 78.5 $m^2$ |
| 3 | 3 ft | 6 ft | 18.84 ft | 28.26 $ft^2$ |
| 4 | 7.5 miles | 15 mi | 47.1 mi | 176.63 $mi^2$ |
| 5 | 5.5 km | 11 km | 34.54 km | 94.99 $km^2$ |
| 6 | 7 cm | 14 cm | 43.96 cm | 153.86 $cm^2$ |
| 7 | 6 ft | 12 ft | 37.68 feet | 113.04 $ft^2$ |
| 8 | 10 m | 20 m | 62.8 m | 314 $m^2$ |
| 9 | 9 in | 18 in | 56.52 in | 254.34 $in^2$ |
| 10 | 4.5 ft | 9 ft | 28.26 ft | 63.585 $ft^2$ |

**Cubes**

1) 12
2) 8 $cm^3$
3) 216 $ft^3$
4) 1,331 $m^3$
5) 2,197 $in^3$
6) 343 $miles^3$
7) 1.728 $km^3$
8) 729 $cm^3$
9) 9.261 $ft^3$
10) 1,728 $mm^3$
11) 0.008 $in^3$
12) 0.001 $km^3$
13) 27
14) 294 $m^2$
15) 150 $ft^2$
16) 121.5 $mm^2$
17) 7.26 $km^2$
18) 726 $cm^2$

**Rectangular Prism**

1) 132 $m^3$
2) 800 $in^3$
3) 600 $m^3$
4) 112 $cm^3$
5) 288 $ft^3$
6) 420 $m^3$
7) 184 $cm^2$
8) 252 $ft^2$
9) 220 $in^2$
10) 438 $m^2$

**Cylinder**

1) 1,004.8 $m^3$
2) 214.6 $cm^3$
3) 9,495.4 $cm^3$
4) 1.1 $m^3$
5) 588.8 $m^3$
6) 452.2 $in^3$
7) 188.4 $m^2$
8) 602.9 $cm^2$
9) 37.7 $cm^2$
10) 401.9 $m^2$

**Pyramids and Cone**

1) 1,600 $yd^3$
2) 1,050 $yd^3$
3) 1,617 $in^3$
4) 392.5 $m^3$
5) 3,014.4 $m^3$
6) 366.33 $cm^3$
7) 1,440 $yd^2$
8) 1,536 $m^2$
9) 678.24 $in^2$
10) 1,205.76 $cm^2$

# ATI TEAS 6 Subject Test – Mathematics

# Chapter 11:
# Statistics and Probability

**Topics that you'll practice in this chapter:**

- ✓ Mean and Median
- ✓ Mode and Range
- ✓ Histograms
- ✓ Stem–and–Leaf Plot
- ✓ Pie Graph
- ✓ Probability Problems

*Mathematics is no more computation than typing is literature.*

*– John Allen Paulos*

**ATI TEAS 6 Subject Test – Mathematics**

## Mean and Median

✒ **Find Mean and Median of the Given Data.**

1) 8, 7, 14, 4, 8

2) 14, 8, 25, 19, 16, 33, 11

3) 23, 18, 15, 12, 17

4) 34, 14, 10, 15, 6, 11

5) 10, 19, 6, 8, 32, 20, 17

6) 17, 26, 39, 69, 20, 6

7) 40, 38, 18, 11, 9, 2, 7, 32, 41

8) 24, 21, 31, 12, 33, 32, 22

9) 16, 14, 20, 41, 15, 20, 38, 4

10) 20, 20, 30, 18, 6, 28, 12, 46

11) 12, 7, 10, 11, 16, 22

12) 10, 29, 27, 12, 2, 15, 10, 3

✒ **Calculate.**

13) In a javelin throw competition, five athletics score 56, 34, 62, 23 and 19 meters. What are their Mean and Median? _____

14) Eva went to shop and bought 8 apples, 14 peaches, 6 bananas, 4 pineapples and 12 melons. What are the Mean and Median of her purchase? _____

15) Bob has 17 black pen, 19 red pen, 14 green pens, 20 blue pens and 5 boxes of yellow pens. If the Mean and Median are 19 respectively, what is the number of yellow pens in each box? _____

**ATI TEAS 6 Subject Test – Mathematics**

# Mode and Range

### ✎ Find Mode and Rage of the Given Data.

1) 4, 3, 7, 3, 3, 4
   Mode: _____   Range: _____

2) 18, 18, 24, 26, 18, 8, 14, 22
   Mode: _____   Range: _____

3) 8, 8, 8, 16, 19, 22, 20, 9, 13
   Mode: _____   Range: _____

4) 24, 24, 14, 28, 20, 18, 20, 24
   Mode: _____   Range: _____

5) 6, 21, 27, 24, 27, 27
   Mode: _____   Range: _____

6) 21, 8, 8, 7, 8, 12, 10, 22, 18, 13
   Mode: _____   Range: _____

7) 7, 4, 4, 6, 13, 13, 13, 0, 2, 2
   Mode: _____   Range: _____

8) 5, 8, 5, 14, 12, 14, 3, 5, 18
   Mode: _____   Range: _____

9) 7, 7, 7, 12, 7, 3, 8, 16, 3, 17
   Mode: _____   Range: _____

10) 15, 15, 19, 16, 4, 16, 10, 15
    Mode: _____   Range: _____

11) 6, 6, 5, 6, 42, 13, 19, 2
    Mode: _____   Range: _____

12) 8, 8, 9, 8, 9, 4, 34, 22
    Mode: _____   Range: _____

### ✎ Calculate.

13) A stationery sold 12 pencils, 56 red pens, 24 blue pens, 20 notebooks, 12 erasers, 21 rulers and 11 color pencils. What are the Mode and Range for the stationery sells?

   Mode: _____   Range: _____

14) In an English test, eight students score 10, 15, 15, 18 18, 16, 15 and 15. What are their Mode and Range? _____

15) What is the range of the first 6 even numbers greater than 8?

   _____

WWW.MathNotion.Com

# ATI TEAS 6 Subject Test – Mathematics

## Times Series

✎ Use the following Graph to complete the table.

| Day | Distance (km) |
|---|---|
| 1 | |
| 2 | |
| | |
| | |
| | |
| | |

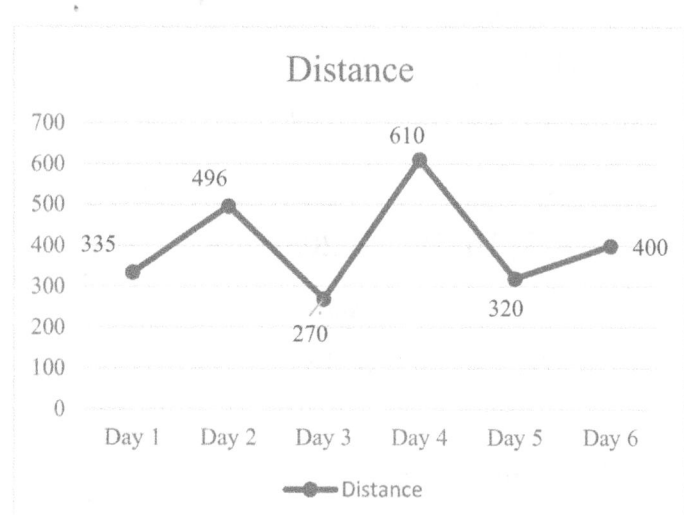

The following table shows the number of births in the US from 2007 to 2012 (in millions).

| Year | Number of births (in millions) |
|---|---|
| 2007 | 4.15 |
| 2008 | 3.70 |
| 2009 | 3.45 |
| 2010 | 3.20 |
| 2011 | 1.75 |
| 2012 | 2.98 |

Draw a Time Series for the table.

WWW.MathNotion.Com

ATI TEAS 6 Subject Test – Mathematics

# Stem–and–Leaf Plot

## ✎ Make stem ad leaf plots for the given data.

1) 24, 26, 29, 20, 53, 27, 51, 55, 36, 21, 37, 30

Stem | Leaf plot

2) 11, 59, 66, 14, 18, 19, 59, 65, 69, 61, 68, 65

Stem | Leaf plot

3) 121, 55, 66, 54, 112, 128, 63, 125, 59, 123, 68, 119

Stem | Leaf plot

4) 51, 32, 100, 56, 84, 36, 107, 56, 85, 39, 56, 106, 89

Stem | Leaf plot

5) 33, 89, 19, 87, 81, 16, 11, 30, 86, 35, 17, 35, 13

Stem | Leaf plot

6) 60, 92, 22, 25, 67, 93, 95, 62, 21, 64, 98, 29

Stem | Leaf plot

WWW.MathNotion.Com

# Pie Graph

The circle graph below shows all Robert's expenses for last month. Robert spent $140 on his hobbies last month.

Answer following questions based on the Pie graph.

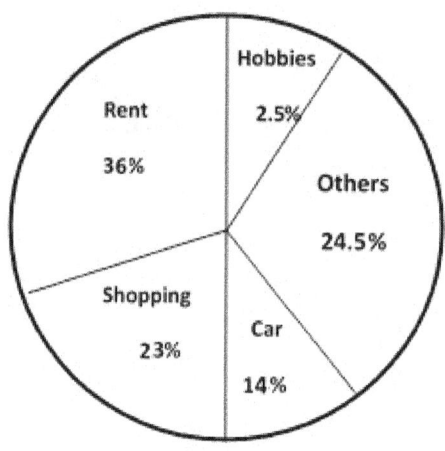

1) How much was Robert's total expenses last month? _____

2) How much did Robert spend on his car last month? _____

3) How much did Robert spend for shopping last month? _____

4) How much did Robert spend on his rent last month? _____

5) What fraction is Robert's expenses for his rent and car out of his total expenses last month? _____

ATI TEAS 6 Subject Test – Mathematics

# Probability Problems

✎ **Calculate.**

1) A number is chosen at random from 1 to 10. Find the probability of selecting number 6 or smaller numbers. _____

2) Bag A contains 18 red marbles and 6 green marbles. Bag B contains 16 black marbles and 8 orange marbles. What is the probability of selecting a green marble at random from bag A? What is the probability of selecting a black marble at random from Bag B? _____

3) A number is chosen at random from 1 to 20. What is the probability of selecting multiples of 4? _____

4) A card is chosen from a well-shuffled deck of 52 cards. What is the probability that the card will be a queen? _____

5) A number is chosen at random from 1 to 15. What is the probability of selecting a multiple of 3 or 5? _____

A spinner numbered 1–8, is spun once. What is the probability of spinning …?

6) an Odd number? _____    7) a multiple of 2? _____

8) a multiple of 5? _____    9) number 10? _____

# ATI TEAS 6 Subject Test – Mathematics

## Answers of Worksheets

### Mean and Median

1) Mean: 8.2, Median: 8
2) Mean: 18, Median: 16
3) Mean: 17, Median: 17
4) Mean: 15, Median: 12.5
5) Mean: 16, Median: 17
6) Mean: 29.5, Median: 23
7) Mean: 22, Median: 18
8) Mean: 25, Median: 24
9) Mean: 21, Median: 18
10) Mean: 22.5, Median: 20
11) Mean: 13, Median: 11.5
12) Mean: 13.5, Median: 11
13) Mean: 38.8, Median: 34
14) Mean: 8.8, Median: 8
15) 5

### Mode and Range

1) Mode: 3, Range: 4
2) Mode: 18, Range: 18
3) Mode: 8, Range: 14
4) Mode: 24, Range: 14
5) Mode: 27, Range: 21
6) Mode: 8, Range: 15
7) Mode: 13, Range: 13
8) Mode: 5, Range: 15
9) Mode: 7, Range: 14
10) Mode: 15, Range: 15
11) Mode: 6, Range: 40
12) Mode: 8, Range: 30
13) Mode: 12, Range: 45
14) Mode: 15, Range: 8
15) 10

### Time series

| Day | Distance (km) |
|-----|---------------|
| 1   | 335           |
| 2   | 496           |
| 3   | 270           |
| 4   | 610           |
| 5   | 320           |
| 6   | 400           |

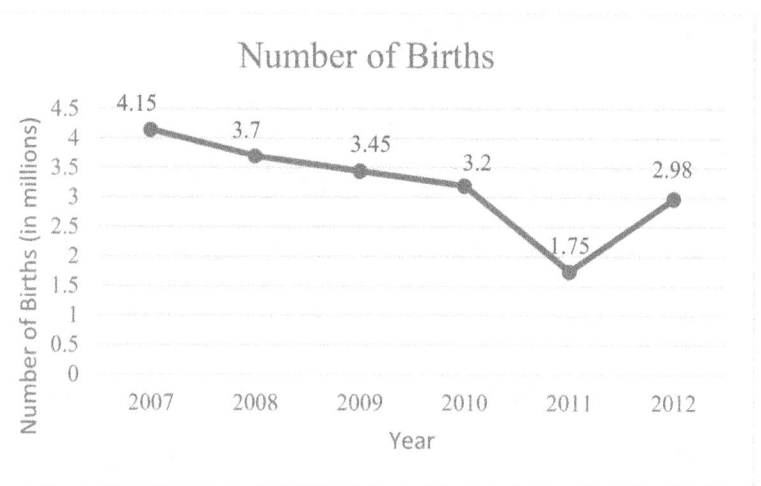

### Stem–And–Leaf Plot

1)

| Stem | leaf     |
|------|----------|
| 2    | 0 1 4 6 7 9 |
| 3    | 0 6 7    |
| 5    | 1 3 5    |

2)

| Stem | leaf      |
|------|-----------|
| 1    | 1 4 8 9   |
| 5    | 9 9       |
| 6    | 1 5 5 6 8 9 |

3)

| Stem | leaf    |
|------|---------|
| 5    | 4 5 9   |
| 6    | 3 6 8   |
| 11   | 2 9     |
| 12   | 1 3 5 8 |

WWW.MathNotion.Com

# ATI TEAS 6 Subject Test – Mathematics

4)

| Stem | leaf |
|---|---|
| 3 | 2 6 9 |
| 5 | 1 6 6 6 |
| 8 | 4 5 9 |
| 10 | 0 6 7 |

5)

| Stem | leaf |
|---|---|
| 1 | 1 3 6 7 9 |
| 3 | 0 3 5 5 |
| 8 | 1 6 7 9 |

6)

| Stem | leaf |
|---|---|
| 2 | 2 1 5 9 |
| 6 | 0 2 4 7 |
| 9 | 2 3 5 8 |

## Pie Graph

1) $5,600

2) $784

3) $1,288

4) $2,016

5) $\frac{1}{2}$

## Probability Problems

1) $\frac{3}{5}$

2) $\frac{1}{4}, \frac{2}{3}$

3) $\frac{1}{4}$

4) $\frac{1}{13}$

5) $\frac{7}{15}$

6) $\frac{1}{2}$

7) $\frac{1}{2}$

8) $\frac{1}{8}$

9) 0

# ATI TEAS 6 Subject Test – Mathematics

ATI TEAS 6 Subject Test – Mathematics

# Chapter 12 : ATI TEAS 6 Test Review

The ATI TEAS (Test of Essential Academic Skills), known as TEAS, is an admissions test for nursing schools, and is designed to assess a student's preparedness entering the health science fields. The last edition (the sixth edition) of the test, called the ATI TEAS 6 Test, was published by ATI Testing on August 31, 2016.

The ATI TEAS 6 Test consists of four multiple-choice sections:

- ✓ **Reading:** 53 Questions – 64 Minutes
- ✓ **Mathematics:** 36 Questions – 54 Minutes
- ✓ **Science:** 53 Questions – 63 Minutes
- ✓ **English and Language Usage:** 28 Questions – 28 Minutes

The Math portion will consist of around 36 multiple-choice questions that address The Math section of the test covers two main topics: Number and Algebra; Measurement and Data.

Students will be allowed to use a four-function calculator during the Math section of the ATI TEAS test. A calculator will be included in the online version and students will be issued one at the testing center during a paper and pencil test.

In this section, there are two complete ATI TEAS 6 Mathematics Tests. Take these tests to see what score you'll be able to receive on a real TEAS test.

# ATI TEAS 6 Subject Test – Mathematics

# Time to Test

**Time to refine your skill with a practice examination**

Take a REAL ATI TEAS 6 Mathematics test to simulate the test day experience. After you've finished, score your test using the answer key.

**Before You Start**

- ✓ You'll need a pencil, a timer, and a four-function calculator to take the test.
- ✓ After you've finished the test, review the answer key to see where you went wrong.
- ✓ Use the answer sheet provided to record your answers. (You can cut it out or photocopy it)
- ✓ You will receive 1 point for every correct answer. There is no penalty for wrong answers.

Good Luck!

# ATI TEAS 6 Subject Test – Mathematics

# ATI TEAS 6 Mathematics Practice Tests Answer Sheets

Remove (or photocopy) these answer sheets and use them to complete the practice tests.

### ATI TEAS 6 Mathematic Practice Test

| # | | # | | # | |
|---|---|---|---|---|---|
| 1 | Ⓐ Ⓑ Ⓒ Ⓓ | 13 | Ⓐ Ⓑ Ⓒ Ⓓ | 25 | Ⓐ Ⓑ Ⓒ Ⓓ |
| 2 | Ⓐ Ⓑ Ⓒ Ⓓ | 14 | Ⓐ Ⓑ Ⓒ Ⓓ | 26 | Ⓐ Ⓑ Ⓒ Ⓓ |
| 3 | Ⓐ Ⓑ Ⓒ Ⓓ | 15 | Ⓐ Ⓑ Ⓒ Ⓓ | 27 | Ⓐ Ⓑ Ⓒ Ⓓ |
| 4 | Ⓐ Ⓑ Ⓒ Ⓓ | 16 | Ⓐ Ⓑ Ⓒ Ⓓ | 28 | Ⓐ Ⓑ Ⓒ Ⓓ |
| 5 | Ⓐ Ⓑ Ⓒ Ⓓ | 17 | Ⓐ Ⓑ Ⓒ Ⓓ | 29 | Ⓐ Ⓑ Ⓒ Ⓓ |
| 6 | Ⓐ Ⓑ Ⓒ Ⓓ | 18 | Ⓐ Ⓑ Ⓒ Ⓓ | 30 | Ⓐ Ⓑ Ⓒ Ⓓ |
| 7 | Ⓐ Ⓑ Ⓒ Ⓓ | 19 | Ⓐ Ⓑ Ⓒ Ⓓ | 31 | Ⓐ Ⓑ Ⓒ Ⓓ |
| 8 | Ⓐ Ⓑ Ⓒ Ⓓ | 20 | Ⓐ Ⓑ Ⓒ Ⓓ | 32 | Ⓐ Ⓑ Ⓒ Ⓓ |
| 9 | Ⓐ Ⓑ Ⓒ Ⓓ | 21 | Ⓐ Ⓑ Ⓒ Ⓓ | 33 | Ⓐ Ⓑ Ⓒ Ⓓ |
| 10 | Ⓐ Ⓑ Ⓒ Ⓓ | 22 | Ⓐ Ⓑ Ⓒ Ⓓ | 34 | Ⓐ Ⓑ Ⓒ Ⓓ |
| 11 | Ⓐ Ⓑ Ⓒ Ⓓ | 23 | Ⓐ Ⓑ Ⓒ Ⓓ | 35 | Ⓐ Ⓑ Ⓒ Ⓓ |
| 12 | Ⓐ Ⓑ Ⓒ Ⓓ | 24 | Ⓐ Ⓑ Ⓒ Ⓓ | 36 | Ⓐ Ⓑ Ⓒ Ⓓ |

WWW.MathNotion.Com

# ATI TEAS 6 Subject Test – Mathematics

ATI TEAS 6 Subject Test – Mathematics

# ATI TEAS 6 Practice Test 1

# Mathematics

- **36 Questions**
- **Total time for this section:** 54 Minutes
- **Calculator is allowed at the test.**

**Administered** *Month Year*

# ATI TEAS 6 Subject Test – Mathematics

1) The width of a garden is 7.96 yards. How many meters is the width of that garden?

   A. 7.28 m

   B. 725.871 m

   C. 52.98 m

   D. 34.352 m

2) The oven temperature reaches 20°C. What's the temperature in degree Fahrenheit?

$$C = \frac{5}{9}(F - 32)$$

   A. 1200° F

   B. 68° F

   C. 72° F

   D. 64° F

3) How many meters is 52,635 centimeters?

   A. 52.6350 m

   B. 5.26350 m

   C. 526.3500 m

   D. 5,263.500 m

# ATI TEAS 6 Subject Test – Mathematics

4) If $x = 3$ what's the value of $4x^2 + 3x - 6$?

   A. 45

   B. 39

   C. 93

   D. 36

5) In seven successive hours, a car travels 26 km, 18 km, 38 km, 28 Km, 48 km and 37 km. In the next six hours, it travels with an average speed of 34 km per hour. Find the total distance the car traveled in 12 hours.

   A. 386 km

   B. 396 km

   C. 399 km

   D. 369 km

6) Find the mean of 345, 262, 272, 398, 392, and 385.

   A. 34.033 …

   B. 342.33 …

   C. 342.5

   D. 340

7) Solve the proportion. $\frac{1.9}{3.4} = \frac{x}{6.8}$

   A. 3. 9

   B. 2.9

   C. 3.8

   D. 2.6

# ATI TEAS 6 Subject Test – Mathematics

8) If $x + y = 12$, what is the value of $4x + 4y$?

   A. 116

   B. 40

   C. 96

   D. 48

9) The equation of a line is given as: $y = -3x + 1$. Which of the following points does not lie on the line?

   A. (4, −11)

   B. (1, –2)

   C. (0, 2)

   D. (−2, 7)

10) If two angles in a triangle measure 42 degrees and 44 degrees, what is the value of the third angle?

    A. 94 Degrees

    B. 86 Degrees

    C. 106 Degrees

    D. 122 Degrees

11) If $8 + x \geq 15$, then $x \geq$ ?

    A. 23

    B. 15

    C. 7

    D. $7x$

# ATI TEAS 6 Subject Test – Mathematics

12) What is the sum of $\frac{7}{18} + \frac{1}{2} + \frac{5}{6}$?

   A. 1.7

   B. 0.16

   C. $3\frac{1}{6}$

   D. 7

13) Last Friday Jacob had $45.86. Over the weekend he received some money for cleaning the attic. He now has $72. How much money did he receive?

   A. $16.41

   B. $26.14

   C. $16.14

   D. $26.41

14) Ella (E) is 11 years older than her friend Ava (A) who is 8 years younger than her sister Sofia (S). If E, A and S denote their ages, which one of the following represents the given information?

   A. $\begin{cases} E = A + 11 \\ S = A - 8 \end{cases}$

   B. $\begin{cases} E = A + 11 \\ A = S + 8 \end{cases}$

   C. $\begin{cases} A = E + 11 \\ S = A - 8 \end{cases}$

   D. $\begin{cases} E = A + 11 \\ A = S - 8 \end{cases}$

# ATI TEAS 6 Subject Test – Mathematics

15) If $x$ is 35% percent of 260, what is $x$?

   A. 96

   B. 91

   C. 225

   D. 85

16) If a rectangle is 29 feet by 46 feet, what is its area?

   A. 1,354

   B. 150

   C. 2,133

   D. 1,334

17) Five years ago, Ann was three times as old as Mia was. If Mia is 12 years old now, how old is Ann?

   A. 26

   B. 18

   C. 16

   D. 36

18) A number is chosen at random from 1 to 10. Find the probability of not selecting a composite number.

   A. $\frac{1}{10}$

   B. $\frac{4}{5}$

   C. $\frac{2}{5}$

   D. $\frac{9}{10}$

ATI TEAS 6 Subject Test – Mathematics

19) What is the value of $x$ in this equation? $7(x+5) = 84$

   A. 5

   B. 15

   C. 7

   D. 12

20) Simplify $\dfrac{\frac{1}{5} - \frac{x+2}{3}}{\frac{x^2}{5} - \frac{4}{5}}$

   A. $\dfrac{-5x+7}{12x^2 - 5}$

   B. $\dfrac{5x - 7}{3x^2 - 12}$

   C. $\dfrac{-5x+7}{5x^2 - 12}$

   D. $\dfrac{-5x - 7}{3x^2 - 12}$

21) Two-kilograms apple and four-kilograms orange cost $41.2. If one-kilogram apple costs $2.6 how much does one-kilogram orange cost?

   A. $9

   B. $9.5

   C. $4

   D. $4.5

# ATI TEAS 6 Subject Test – Mathematics

22) The average weight of 14 girls in a class is 55 kg and the average weight of 26 boys in the same class is 65 kg. What is the average weight of all the 40 students in that class?

   A. 56

   B. 62

   C. 64.3

   D. 61.5

23) A circle has a diameter of 2.8 inches. What is its approximate circumference?

   A. 6

   B. 9

   C. 7

   D. 12

24) In a certain bookshelf of a library, there are 50 biology books, 60 history books, and 90 language books. What is the ratio of the number of biology books to the total number of books in this bookshelf?

   A. $\frac{1}{2}$

   B. $\frac{1}{4}$

   C. $\frac{3}{4}$

   D. $\frac{4}{5}$

# ATI TEAS 6 Subject Test – Mathematics

25) The circumference of a circle is 36cm. what is the approximate radius of the circle?

   A. 5.1 cm

   B. 5.7 cm

   C. 7.5 cm

   D. 7.1 cm

**Questions 26 and 27 are based on following chart.**

The following pie chart shows the expenses of Mr. Janson's family in December. The total expenses in December was $6,100.

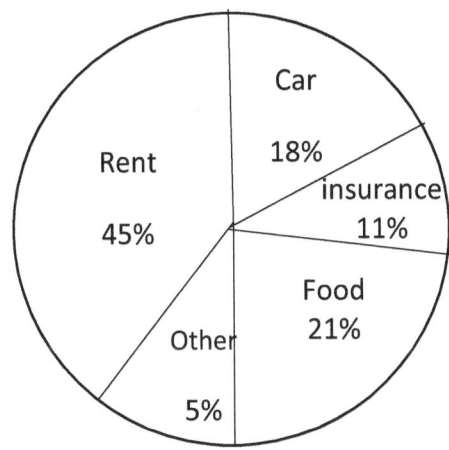

26) What percent of the expenses goes for other and rent combined?

   A. 50 %

   B. 40 %

   C. 46 %

   D. 46 %

# ATI TEAS 6 Subject Test – Mathematics

27) How much did Mr. Janson's family spend on insurance?

   A. $266

   B. $671

   C. $872

   D. $1,671

28) A football team won exactly 60% of the games it played during last session. Which of the following could be the total number of games the team played last season?

   A. 32

   B. 37

   C. 25

   D. 19

29) I've got 64 quarts of milk and my family drinks 4 gallons of milk per week. How many weeks will that last us?

   A. 6 Weeks

   B. 4.2 Weeks

   C. 4.4 Weeks

   D. 4 Weeks

# ATI TEAS 6 Subject Test – Mathematics

30) 8 liters of water are poured into an aquarium that's 16 cm long, 5 cm wide. How many cm will the water level in the aquarium rise due to this added water? (1 liter of water = 1,000 cm$^3$)

   A. 1,000

   B. 100

   C. 80

   D. 21

31) At a Zoo, the ratio of lions to tigers is 3 to 2. Which of the following could NOT be the total number of lions and tigers in the zoo?

   A. 70

   B. 25

   C. 77

   D. 100

32) In a bundle of 60 pencils, 21 are red and the rest are blue. What percent of the bundle is composed of blue pencils?

   A. 6.2%

   B. 62%

   C. 65%

   D. 6.5%

# ATI TEAS 6 Subject Test – Mathematics

33) If $a$ inches of rain fall in four minutes, how many inches will fall in $b$ hours?

   A. $15\frac{a}{b}$

   B. $15a$

   C. $15b$

   D. $15ab$

34) A card is drawn at random from a standard 52–card deck, what is the probability that the card is of Hearts or Diamonds?

   A. $\frac{3}{4}$

   B. $\frac{1}{5}$

   C. $\frac{5}{13}$

   D. $\frac{1}{2}$

35) What is the value of the expression $3(2x + 5y) + (8 - 2x)^2$ when $x = 3$ and y= −1 ?

   A. 7

   B. 17

   C. 21

   D. 14

36) If a gas tank can hold 25 gallons, how many gallons does it contain when it is $\frac{4}{5}$ full?

A. 25

B. 20

C. 100

D. 30

# ATI TEAS 6 Subject Test – Mathematics

# ATI TEAS 6 Practice Test 2

# Mathematics

- 36 Questions
- **Total time for this section:** 54 Minutes
- Calculator is allowed at the test.

**Administered** *Month Year*

# ATI TEAS 6 Subject Test – Mathematics

1) In the simplest form, $\frac{16}{28}$ is

   A. $\frac{1}{7}$

   B. $\frac{7}{3}$

   C. $\frac{7}{4}$

   D. $\frac{4}{7}$

2) $\frac{(31\,feet + 3\,yards)}{8} = \underline{\phantom{xxx}}$

   A. 34 feet

   B. 45 feet

   C. 15 feet

   D. 5 feet

3) The sum of 9 numbers is greater than 630 and less than 720. Which of the following could be the average (arithmetic mean) of the numbers?

   A. 55

   B. 60

   C. 75

   D. 100

4) If $x = 5$, then $\frac{5^5}{x} =$

   A. 625

   B. 3,125

   C. 1,225

   D. 25

# ATI TEAS 6 Subject Test – Mathematics

5) The distance between cities A and B is approximately 2,700 miles. If you drive an average of 35 miles per hour, how many hours will it take you to drive from city A to city B?

   A. approximately 87 hours

   B. approximately 77 hours

   C. approximately 97 hours

   D. approximately 46 hours

6) A swimming pool holds 2,310 cubic feet of water. The swimming pool is 55 feet long and 7 feet wide. How deep is the swimming pool?

   A. 12

   B. 10

   C. 6

   D. 4

7) Chris is 22 miles ahead of Joe running at 3.5 miles per hour and Joe is running at the speed of 9 miles per hour. How long does it take Joe to catch Chris?

   A. 5 hours

   B. 4 hours

   C. 8 hours

   D. 10 hours

# ATI TEAS 6 Subject Test – Mathematics

8) A bread recipe calls for $6\frac{5}{6}$ cups of flour. If you only have $3\frac{1}{6}$ cups, how much more flour is needed?

   A. 3

   B. $\frac{11}{3}$

   C. 6

   D. $\frac{1}{6}$

9) The perimeter of a rectangular yard is 180 meters. What is its length if its width is twice its length?

   A. 30 meters

   B. 10 meters

   C. 25 meters

   D. 35 meters

10) If $8.6 < x \leq 11.0$, then $x$ cannot be equal to:

   A. 8.6

   B. 10

   C. 9.6

   D. 10.5

# ATI TEAS 6 Subject Test – Mathematics

11) If $(6.4 + 9.4 + 3.2)\ x = x$, then what is the value of $x$?

   A. 3

   B. $\frac{1}{9}$

   C. 19

   D. 0

12) The equation of a line is given as: $y = 7x - 12$. Which of the following points does not lie on the line?

   A. (2, 2)

   B. (–2, –26)

   C. (3, 11)

   D. (0, –12)

13) If $a = 8$ what is the value of $b$ in the following equation?

$$b = \frac{a^2}{4} + c$$

   A. $4 + c$

   B. $8 + c$

   C. $32 + c$

   D. $16 + c$

# ATI TEAS 6 Subject Test – Mathematics

14) The sum of two numbers is $x$. If one of the numbers is 7, then three times the other number would be?

   A. $7x$

   B. $7 + x \times 2$

   C. $3(x + 7)$

   D. $3(x - 7)$

15) If $x = \frac{7}{3}$ then $\frac{1}{x} = ?$

   A. $\frac{7}{3}$

   B. $\frac{3}{7}$

   C. 7

   D. 3

16) Which of the following is the product of $3\frac{1}{4}$ and $3\frac{3}{7}$?

   A. $11\frac{1}{28}$

   B. $11\frac{1}{7}$

   C. $7\frac{1}{7}$

   D. $7\frac{1}{11}$

# ATI TEAS 6 Subject Test – Mathematics

17) How many $\frac{1}{9}$ pound paperback books together weigh 80 pounds?

    A. 38

    B. 80

    C. 720

    D. 740

18) 6 feet, 12 inches +4 feet, 15 inches equals to how many inches?

    A. 147 inches

    B. 174 inches

    C. 63 inches

    D. 84 inches

19) $\frac{3}{6}$ is equals to:

    A. 0.5

    B. 0.3

    C. 0.05

    D. 0.03

ATI TEAS 6 Subject Test – Mathematics

**Questions 20 and 21 are based on following Pie Chart.**

The following pie chart shows the time Elise spent to work on his homework last week.

The total time Elise spent on his homework last week was 80 hours.

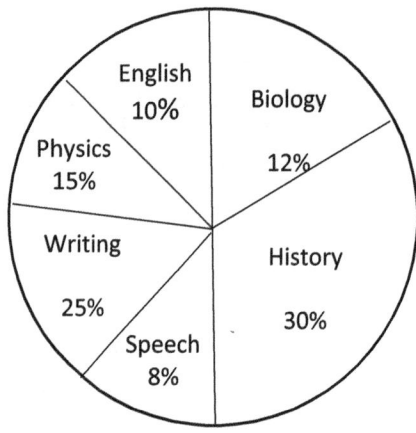

20) How much time did Elise spend on English last week?

　A. 4.5 hours

　B. 6 hours

　C. 9.5 hours

　D. 8 hours.

21) What many hours did Elise spend doing the Writing and Physics?

　A. 18 hours

　B. 80 hours

　C. 12 hours

　D. 15 hours

# ATI TEAS 6 Subject Test – Mathematics

22) A circle has a diameter of 6 inches. What is its approximate circumference?

   A. 33.64

   B. 18.84

   C. 12.56

   D. 3.14

23) If a circle has a radius of 15 feet, what's the closest approximation of its circumference?

   A. 94

   B. 47

   C. 96

   D. 706

24) If two angles in a triangle measure 53 degrees and 43 degrees, what is the value of the third angle?

   A. 96 degrees

   B. 84 degrees

   C. 42 degrees

   D. 36 degrees

25) What is the sum of $\frac{4}{5} + \frac{1}{10} + \frac{3}{4}$?

   A. 0.65

   B. 3

   C. $1\frac{13}{20}$

   D. $1\frac{17}{10}$

# ATI TEAS 6 Subject Test – Mathematics

26) Which of the following graphs represents the compound inequality?

$$-3 \leq 5x - 8 < 22$$

A.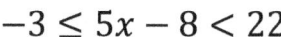

B.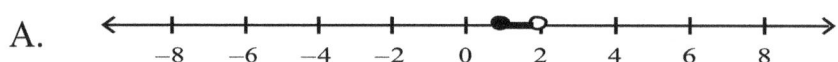

C.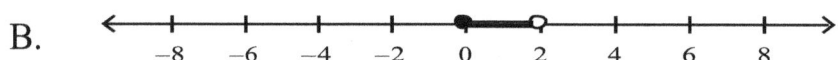

D.

27) What is the equivalent temperature of 113°F in Celsius?

$$C = \frac{5}{9}(F - 32)$$

A. 40

B. 65

C. 45

D. 55

28) Julie gives 7 pieces of candy to each of her friends. If Julie gives all her candy away, which amount of candy could have been the amount she distributed?

A. 184

B. 536

C. 294

D. 382

# ATI TEAS 6 Subject Test – Mathematics

29) With an 32% discount, Ella was able to save $45.73 on a dress. What was the original price of the dress?

   A. $67.25

   B. $142.9

   C. $124.9

   D. $76.25

30) The letters represent two decimals listed below. One of the decimals is equivalent to $\frac{1}{125}$ and the other is equivalent to $\frac{1}{25}$. What is the product of C and D?

   $\qquad$ 0.ABC  $\quad$  0.0D

   A. 8

   B. 0

   C. 16

   D. 32

31) The marked price of a computer is D dollar. Its price decreased by 30% in January and later increased by 25% in February. What is the final price of the computer in D dollar?

   A. 0.785 D

   B. 0.875 D

   C. 0.825 D

   D. 8.25 D

## ATI TEAS 6 Subject Test – Mathematics

32) If five times a number added to 9 equals to 54, what is the number?

   A. 18

   B. 11

   C. 10

   D. 9

33) A bank is offering 3.5% simple interest on a savings account. If you deposit $22,000, how much interest will you earn in five years?

   A. $850

   B. $3,850

   C. $770

   D. $2,695

34) A football team had $45,000 to spend on supplies. The team spent $20,000 on new balls. New sport shoes cost $220 each. Which of the following inequalities represent how many new shoes the team can purchase?

   A. $220x + 20{,}000 \leq 45{,}000$

   B. $220x + 20{,}000 \geq 45{,}000$

   C. $20{,}000x + 220 \leq 45{,}000$

   D. $20{,}000x + 220 \geq 45{,}000$

# ATI TEAS 6 Subject Test – Mathematics

35) What is the value of $x$ in the following equation?

$$\frac{3}{7}x + \frac{1}{4} = \frac{1}{2}$$

A. 7

B. $\frac{1}{12}$

C. $\frac{7}{12}$

D. $\frac{3}{7}$

36) Two dice are thrown simultaneously, what is the probability of getting a sum of 5 or 9?

A. $\frac{3}{18}$

B. $\frac{3}{8}$

C. $\frac{2}{9}$

D. $\frac{7}{36}$

# ATI TEAS 6 Subject Test – Mathematics

# Chapter 13 : Answers and Explanations

## Answer Key

❋ Now, it's time to review your results to see where you went wrong and what areas you need to improve!

### ATI TEAS 6 - Mathematics

| Practice Test - 1 | | | | Practice Test - 2 | | | |
|---|---|---|---|---|---|---|---|
| 1 | A | 19 | C | 1 | D | 19 | A |
| 2 | B | 20 | D | 2 | D | 20 | D |
| 3 | C | 21 | A | 3 | C | 21 | C |
| 4 | B | 22 | D | 4 | A | 22 | B |
| 5 | C | 23 | B | 5 | B | 23 | A |
| 6 | B | 24 | B | 6 | C | 24 | B |
| 7 | C | 25 | B | 7 | B | 25 | C |
| 8 | D | 26 | A | 8 | B | 26 | D |
| 9 | C | 27 | B | 9 | A | 27 | C |
| 10 | A | 28 | C | 10 | A | 28 | C |
| 11 | C | 29 | D | 11 | D | 29 | B |
| 12 | A | 30 | B | 12 | C | 30 | D |
| 13 | B | 31 | C | 13 | D | 31 | B |
| 14 | D | 32 | C | 14 | D | 32 | D |
| 15 | B | 33 | D | 15 | B | 33 | B |
| 16 | D | 34 | D | 16 | B | 34 | A |
| 17 | A | 35 | A | 17 | C | 35 | C |
| 18 | C | 36 | B | 18 | A | 36 | C |

# ATI TEAS 6 Subject Test – Mathematics

ATI TEAS 6 Subject Test – Mathematics

# Answers and Explanations
# ATI TEAS 6 - Mathematics
# Practice Test 1

**1) Answer: A**

$m = \frac{yd}{1.0936} \rightarrow m = \frac{7.96}{1.0936} = 7.27125 \cong 7.28$

**2) Answer: B**

Plug in 20 for C in the equation: $20 = \frac{5}{9}(F - 32)$

$180 = 5F - 160$

$180 + 160 = 5F \Rightarrow \frac{180 + 160}{5} = F$

$\frac{340}{5} = F \Rightarrow F = 68$

**3) Answer: C**

1 meter = 100 centimeters

$52{,}635 \times 0.01 = 526.35$

**4) Answer: B**

$4x^2 + 3x - 6 = 4(3)^2 + 3(3) - 6 = 36 + 9 - 6 = 39$

**5) Answer: C**

Add the first 6 numbers. $26 + 18 + 38 + 28 + 48 + 37 = 195$

To find the distance traveled in the next 6 hours, multiply the average by number of hours.

Distance = Average × Rate = $34 \times 6 = 204$

Add both numbers. $195 + 204 = 399$

**6) Answer: B**

Mean: $\frac{\text{sum of the data}}{\text{of data entires}} = \frac{345 + 262 + 272 + 398 + 392 + 385}{6} = \frac{2{,}054}{6} = 342.33$

**7) Answer: C**

$\frac{1.9}{3.4} = \frac{x}{6.8} \rightarrow x = \frac{1.9 \times 6.8}{3.4} = \frac{12.92}{3.4} = 3.8$

# ATI TEAS 6 Subject Test – Mathematics

**8) Answer: D**

$x + y = 12$ Then: $4x + 4y = 12 \times 4 = 48$

**9) Answer: C**

Let's review the choices provided. Put the values of $x$ and $y$ in the equation.

A. $(4, -11) \Rightarrow x = 4 \Rightarrow y = -11$    This is true!

B. $(1, -2) \Rightarrow x = 1 \Rightarrow y = -2$    This is true!

C. $(0, 2) \Rightarrow x = 0 \Rightarrow y = 1$    This is not true!

D. $(-2, 7) \Rightarrow x = -2 \Rightarrow y = 7$    This is true!

**10) Answer: A**

All angles in a triangle sum up to 180 degrees. Two angles add up to 90 degrees.

$42° + 44° = 86°$, then the third angle is: $180° - 86° = 94°$

**11) Answer: C**

$8 + x \geq 15, \quad x \geq 15 - 8, \quad x \geq 7$

**12) Answer: A**

$\frac{7}{18} + \frac{1}{2} + \frac{5}{6} = \frac{7+9+15}{18} = \frac{31}{18} = 1.722 = 1.7$

**13) Answer: B**

$72 - $45.86 = 26.14

**14) Answer: D**

Choice D represents the given information.

$E = 11 + A, \quad A = S - 8$

**15) Answer: B**

$\frac{35}{100} \times 260 = 91$

**16) Answer: D**

Area of a rectangle = width × length = $29 \times 46 = 1,334$

**17) Answer: A**

Five years ago, Ann was three times as old as Mia. Mia is 12 years now. Therefore, 5 years ago Mia was 7 years. five years ago, Ann was: $A = 7 \times 3 = 21$

Now Ann is 26 years old: $21 + 5 = 26$

# ATI TEAS 6 Subject Test – Mathematics

**18) Answer: C**

The list of composite numbers from 1 to 10 is: 2, 4, 6, 8, 9, 10

There are 6 numbers in the list. Therefore, the probability of selecting a composite number is 6 out of 10 and the probability of not selecting a composite number is 4 out of 10 or ($\frac{2}{5}$).

**19) Answer: C**

Solve for $x$ in the equation.

$7(x + 5) = 84 \rightarrow 7x + 35 = 84 \rightarrow 7x = 84 - 35 = 49 \rightarrow x = 49 \div 7 = 7$

**20) Answer: D**

Simplify: $\dfrac{\frac{1}{5} - \frac{x+2}{3}}{\frac{x^2}{5} - \frac{4}{5}} = \dfrac{\frac{1}{5} - \frac{x+2}{3}}{\frac{x^2-4}{5}} = \dfrac{\frac{3 - 5x - 10}{15}}{\frac{x^2-4}{5}} \Rightarrow \dfrac{-5x-7}{15} \div \dfrac{x^2-4}{5}$

Then: $\dfrac{-5x-7}{15} \times \dfrac{5}{x^2-4} = \dfrac{5(-5x-7)}{15(x^2-4)} = \dfrac{-5x-7}{3(x^2-4)} = \dfrac{-5x-7}{3x^2-12}$

**21) Answer: A**

Let $x$ be one-kilogram orange cost, then: $4x + (2 \times 2.6) = 41.2 \rightarrow 4x + 5.2 = 41.2 \rightarrow 4x = 41.2 - 5.2 \rightarrow 4x = 36 \rightarrow x = \dfrac{36}{4} = \$9$

**22) Answer: D**

$\text{average} = \dfrac{\text{sum of terms}}{\text{number of terms}}$

The sum of the weight of all girls is: $14 \times 55 = 770$ kg

The sum of the weight of all boys is: $26 \times 65 = 1,690$ kg

The sum of the weight of all students is: $770 + 1,690 = 2,460$ kg

Average $= \dfrac{2,460}{40} = 61.5$

**23) Answer: B**

Diameter $= 2r \Rightarrow 2.8 = 2r \Rightarrow r = 1.4$

Circumference $= 2\pi r \Rightarrow C = 2\pi(1.4) \Rightarrow C = 2.8\pi = 8.792 \cong 9$

**24) Answer: B**

Number of biology book: 50, total number of books; $50 + 60 + 90 = 200$

# ATI TEAS 6 Subject Test – Mathematics

the ratio of the number of biology books to the total number of books is: $\frac{50}{200}=\frac{1}{4}$

**25) Answer: B**

$C = 2\pi r \Longrightarrow 36 = 2\pi r$

$r = \frac{36}{2\pi} = 5.732 \cong 5.7$

**26) Answer: A**

$45\% + 5\% = 50\%$

**27) Answer: B**

$\$6,100 \times \frac{11}{100} = \$671$

**28) Answer: C**

Choices A, B, and D are incorrect because 60% of each of the numbers is a non-whole number.

   A. 32       60% of 32 = 0.60 × 32 = 19.2

   B. 37       60% of 37 = 0.60 × 37 = 22.2

   C. 25       60% of 25 = 0.60 × 25 = 15

   D. 19       60% of 19 = 0.60 × 19 = 11.4

**29) Answer: D**

1 quart = 0.25 gallon

64 quarts = 64 × 0.25 = 16 gallons,

then: $\frac{16}{4} = 4$ weeks

**30) Answer: B**

*One liter* = 1,000 cm³ → 8 *liters* = 8,000 cm³

$8,000 = 16 \times 5 \times h \rightarrow h = \frac{8,000}{80} = 100$ cm

**31) Answer: C**

The ratio of lions to tigers is 3 to 2 at the zoo. Therefore, total number of lions and tigers must be divisible by 5.

3 + 2 = 5

From the numbers provided, only 77 is not divisible by 5.

# ATI TEAS 6 Subject Test – Mathematics

**32) Answer: C**

$60 - 21 = 39$

$\frac{39}{60} = 0.65$

$0.65 \times 100\% = 65\%$

**33) Answer: D**

(hour: $\frac{4}{60} = \frac{1}{15}$); $\frac{a}{\frac{1}{15}} = \frac{x}{b} \rightarrow x = \frac{ab}{\frac{1}{15}} = 15ab$

**34) Answer: D**

The probability of choosing a heart or diamonds is $\frac{26}{52} = \frac{1}{2}$

**35) Answer: A**

$3(2x + 5y) + (8 - 2x)^2 = 3(2(3) + 5(-1)) + (8 - 2(3))^2 = 3(1) + (2)^2 = 3 + 4 = 7$

**36) Answer: B**

$\frac{4}{5} \times 25 = \frac{100}{5} = 20$

# ATI TEAS 6 Subject Test – Mathematics

# Answers and Explanations
# ATI TEAS 6 - Mathematics
# Practice Test 2

**1) Answer: D**

$\frac{16}{28} = \frac{4}{7}$

**2) Answer: D**

$1\ yard = 3\ feet$

$\frac{(31\ feet + 3\ yards)}{8} = \frac{(31\ feet + 9\ feet)}{8} = \frac{(40\ feet)}{8} = 5\ feet$

**3) Answer: C**

Let $x$ be the average of numbers. Then:

$\frac{630}{9} < x < \frac{720}{9}$

$70 < x < 80$

From choices provided, only 75 is correct.

**4) Answer: A**

$\frac{5^5}{5} = 5^4 = 625$

**5) Answer: B**

$\text{Speed} = \frac{distance}{time}$

$35 = \frac{2{,}700}{time} \rightarrow time = \frac{2{,}700}{35} = 77.14 \cong 77$

**6) Answer: C**

Volume = length × width × height

$2{,}310 = 55 \times 7 \times height \rightarrow height = 6$

**7) Answer: B**

The distance between Chris and Joe is 22 miles. Chris running at 3.5 miles per hour and Joe is running at the speed of 9 miles per hour. Therefore, every hour the distance is 5.5 miles less.

$22 \div 5.5 = 4$

# ATI TEAS 6 Subject Test – Mathematics

**8) Answer: B**

$6\frac{5}{6} - 3\frac{1}{6} = 6 + \frac{5}{6} - 3 - \frac{1}{6} = 3\frac{4}{6} = 3\frac{2}{3} = \frac{11}{3}$

**9) Answer: A**

Perimeter of a rectangle $= 2(width + length)$

$P = 180, width = 2 \times length$

Then: $180 = 2(2length + length) \rightarrow 180 = 6length \rightarrow length = 30$

**10) Answer: A**

If $8.6 < x \leq 11.0$, then $x$ cannot be equal to 8.6.

**11) Answer: D**

$(6.4 + 9.4 + 3.2)\,x = x$

$19x = x$

Then $x = 0$

**12) Answer: C**

Let's review the choices provided. Put the values of $x$ and $y$ in the equation.

A. (2, 2) ⇒ $x = 2$ ⇒ $y = 2$     This is true!

B. (−2, −26) ⇒ $x = -2$ ⇒ $y = -26$ This is true!

C. (3, 11)      ⇒ $x = 3$ ⇒ $y = 9$     This is not true!

D. (0, −12) ⇒ $x = 0$ ⇒ $y = -12$ This is true!

Only choice C does not work in the equation.

**13) Answer: D**

If $a = 8$ then: $b = \frac{8^2}{4} + c \Rightarrow b = \frac{8^2}{4} + c = 16 + c$

**14) Answer: D**

Let $a$ and $b$ be the numbers. Then: $a + b = x$

$a = 7 \rightarrow 7 + b = x \rightarrow b = x - 7$

$3b = 3(x - 7)$

**15) Answer: B**

$\frac{1}{x} = \frac{\frac{1}{1}}{\frac{7}{3}} = \frac{3}{7}$

# ATI TEAS 6 Subject Test – Mathematics

**16) Answer: B**

Converting mixed numbers to fractions, our initial equation becomes
$\frac{13}{4} \times \frac{24}{7}$, Applying the fractions formula for multiplication
$\frac{13 \times 24}{4 \times 7} = \frac{312}{28} = 11\frac{1}{7}$

**17) Answer: C**

$80 \div \frac{1}{9} = 720$

**18) Answer: A**

1 foot = 12 inches

6 feet, 12 inches = 84 inches

4 feet, 15 inches = 63 inches

84 + 63 = 147

**19) Answer: A**

$\frac{3}{6} = 0.5$

**20) Answer: D**

Elise spent 10% of his total time (80 hours) on English. Then: $\frac{10}{100} \times 80 = 8$

**21) Answer: C**

Elise spent 15% of his time on Writing and Physics. Then: $\frac{15}{100} \times 80 = 12$

**22) Answer: B**

Circumference = $2\pi r$

C = $2\pi \times 3 = 6\pi$; $\pi = 3.14 \rightarrow$ C = $6\pi$ = 18.84

**23) Answer: A**

Circumference = $2\pi r \rightarrow$ Circumference = 2(3.14)(15) = 94.2 $\cong$ 94

**24) Answer: B**

All angles in a triangle sum up to 180 degrees.

53 + 43 = 96

180 − 96 = 84,    The third angle is 84 degrees.

# ATI TEAS 6 Subject Test – Mathematics

**25) Answer: C**

$$\frac{4}{5} + \frac{1}{10} + \frac{3}{4} = \frac{16+2+15}{20} = \frac{33}{20} = 1\frac{13}{20} = 1.65$$

**26) Answer: D**

Solve for $x$.

$-3 \leq 5x - 8 < 22 \Rightarrow$ (add 8 all sides) $-3 + 8 \leq 5x - 8 + 8 < 22 + 8 \Rightarrow 5 \leq 5x < 30 \Rightarrow$ (divide all sides by 5) $1 \leq x < 6$

$x$ is between 1 and 6. Choice D represent this inequality.

**27) Answer: C**

Plug in 113 for F in the equation:

$C = \frac{5}{9}(F - 32) = \frac{5}{9}(113 - 32) = \frac{5}{9}(81) = 45$

**28) Answer: C**

Since Julie gives 7 pieces of candy to each of her friends, then, then number of pieces of candies must be divisible by 7.

A. $184 \div 7 = 22.29$

B. $536 \div 7 = 76.57$

C. $294 \div 7 = 42$

D. $382 \div 7 = 54.57$

Only choice C gives a whole number.

**29) Answer: B**

$32\%$ of $x = 45.73 \rightarrow x = \frac{100 \times 45.73}{32} \cong 142.9$

**30) Answer: D**

$\frac{1}{125} = 0.008 \rightarrow C = 8, \quad \frac{1}{25} = 0.04 \rightarrow D = 4 \rightarrow C \times D = 8 \times 4 = 32$

**31) Answer: B**

To find the discount, multiply the number by (100% – rate of discount).

Therefore, for the first discount we get: (D) (100% – 30%) = (D) (0.70) = 0.70 D

For increase of 25%: (0.70 D) (100% + 25%) = (0.70 D) (1.25) = 0.875 D = 87.5% of D or 0.875D

# ATI TEAS 6 Subject Test – Mathematics

**32) Answer: D**

Let $x$ be the number. Then: $5x + 9 = 54$

Solve for $x$: $5x + 9 = 54 \rightarrow 5x = 54 - 9 = 45 \rightarrow x = 45 \div 5 = 9$

**33) Answer: B**

Use simple interest formula:

$I = prt$ (I = interest, p = principal, r = rate, t = time)

$I = (22,000)(0.035)(5) = 3,850$

**34) Answer: A**

Let $x$ be the number of new shoes the team can purchase. Therefore, the team can purchase $220\ x$.

The team had $45,000 and spent $20,000. Now the team can spend on new shoes $20,000 at most.

Now, write the inequality:

$220x + 20,000 \leq 45,000$

**35) Answer: C**

$\frac{3}{7}x + \frac{1}{4} = \frac{1}{2} \rightarrow \frac{3}{7}x = \frac{1}{2} - \frac{1}{4} \rightarrow \frac{3}{7}x = \frac{2-1}{4} \rightarrow \frac{3}{7}x = \frac{1}{4} \rightarrow x = \frac{7}{12}$

**36) Answer: C**

To get a sum of 5 for two dice, we can get 4 different options:

(1, 4), (4, 1), (2, 3), (3, 2)

To get a sum of 9 for two dice, we can get 4 different options:

(3, 6), (6, 3), (4, 5), (5, 4)

Therefore, there are 8 options to get the sum of 5 or 9.

Since, we have $6 \times 6 = 36$ total options, the probability of getting a sum of 5 or 9 is 8 out of 36 or $\frac{8}{36} = \frac{2}{9}$.

## "End"

www.ingramcontent.com/pod-product-compliance
Lightning Source LLC
Chambersburg PA
CBHW080438110426
42743CB00016B/3207